VOLUME TWO HUNDRED AND FOUR

ADVANCES IN IMAGING AND ELECTRON PHYSICS

VOLUME TWO HUNDRED AND FOUR

ADVANCES IN IMAGING AND ELECTRON PHYSICS

Edited by

PETER W. HAWKES
CEMES-CNRS
Toulouse, France

Cover photo credit:
The cover picture is taken from Fig. 12 of the chapter by Román Castañeda and Giorgio Matteucci (p. 34).

Academic Press is an imprint of Elsevier
125 London Wall, London EC2Y 5AS, United Kingdom
525 B Street, Suite 1800, San Diego, CA 92101-4495, United States
50 Hampshire Street, 5th Floor, Cambridge, MA 02139, United States
The Boulevard, Langford Lane, Kidlington, Oxford OX5 1GB, United Kingdom

ISBN: 978-0-12-812086-6
ISSN: 1076-5670

For information on all Academic Press publications
visit our website at https://www.elsevier.com/books-and-journals

Publisher: Zoe Kruze
Acquisition Editor: Jason Mitchell
Editorial Project Manager: Shellie Bryant
Production Project Manager: Divya Krishna Kumar
Designer: Matthew Limbert

Typeset by VTeX

CONTENTS

CONTRIBUTORS

Román Castañeda
Physics School, Universidad Nacional de Colombia Sede Medellín, Medellín, Colombia

Taryl L. Kirk
Rowan University, Glassboro, NJ, USA
Educational Testing Service, Princeton, NJ, USA

Giorgio Matteucci
University of Bologna, Bologna, Italy

Allen J.F. Metherell
Formerly Cavendish Laboratory, Cambridge, England

Jos B.T.M. Roerdink
University of Groningen, Groningen, The Netherlands

Jasper van de Gronde
University of Groningen, Groningen, The Netherlands

PREFACE

In parallel with reviews of progress in established subjects, these Advances occasionally include more revolutionary or at least very new material. Thus the first full-length account of the scanning electron microscope was published here in 1965, the year that the first commercial instrument appeared on the market. In a different area, the ideas of Henning Harmuth, which were at first difficult to place in regular journals, were presented here at length. I am therefore delighted to open this volume with the account of a new physical principle for the interference of light and material particles by Román Castañeda and Giorgio Matteucci. This makes the ambitious and in my opinion, justified claim of unifying the wave and particle pictures of interference phenomena in a way that has not been achieved before. I shall not attempt to summarize it here but I anticipate that it will be widely read.

The second chapter is an account by Taryl Kirk of the near-field emission mode of scanning electron microscopy. A different version of this chapter was withdrawn from an earlier volume and the present version should be consulted for details of this original mode of operation of the scanning electron microscope.

The third chapter brings us to a regular feature of the series, mathematical morphology. Here, Jasper van de Gronde and Jos B.T.M. Roerdink suggest ways of using mathematical morphology to study objects that do not fit naturally into the familiar theory. Multidimensional data are an example of such objects and the authors explain how sponges, a generalization of lattices, can be helpful. Other types on non-traditional data are examined and ways of incorporating them are discussed.

We conclude with another article from *Advances in Optical and Electron Microscopy*, which certainly deserves to be made easily accessible. This is a long account by Allen Metherell of energy analyzing and energy selecting electron microscopes, with very full studies of such devices as the Castaing–Henry analyzer and the magnetic analyzer introduced by Ichinokawa. It is lavishly illustrated with simple line diagrams and photographs and is essential reading for anyone interested in analytical electron microscope instrumentation.

As always, I am very grateful to the authors for their ability to make unusual material readable by the non-specialist.

Peter W. Hawkes

FUTURE CONTRIBUTIONS

S. Ando
Gradient operators and edge and corner detection

D. Batchelor
Soft x-ray microscopy

E. Bayro Corrochano
Quaternion wavelet transforms

C. Beeli
Structure and microscopy of quasicrystals

C. Bobisch, R. Möller
Ballistic electron microscopy

F. Bociort
Saddle-point methods in lens design

K. Bredies
Diffusion tensor imaging

A. Broers
A retrospective

A. Cornejo Rodriguez, F. Granados Agustin
Ronchigram quantification

J. Elorza
Fuzzy operators

R.G. Forbes
Liquid metal ion sources

P.L. Gai, E.D. Boyes
Aberration-corrected environmental microscopy

S. Golodetz
Watersheds and waterfalls

R. Herring, B. McMorran
Electron vortex beams

F. Houdellier, A. Arbouet
Ultrafast electron microscopy

M.S. Isaacson
Early STEM development

K. Ishizuka
Contrast transfer and crystal images

K. Jensen, D. Shiffler, J. Luginsland
Physics of field emission cold cathodes

U. Kaiser
The sub-Ångström low-voltage electron microscope project (SALVE)

K. Kimoto
Monochromators for the electron microscope

O.L. Krivanek
Aberration-corrected STEM

M. Kroupa
The Timepix detector and its applications

B. Lencová
Modern developments in electron optical calculations

H. Lichte
Developments in electron holography

M. Matsuya
Calculation of aberration coefficients using Lie algebra

J.A. Monsoriu
Fractal zone plates

L. Muray
Miniature electron optics and applications

M.A. O'Keefe
Electron image simulation

V. Ortalan
Ultrafast electron microscopy

D. Paganin, T. Gureyev, K. Pavlov
Intensity-linear methods in inverse imaging

N. Papamarkos, A. Kesidis
The inverse Hough transform

H. Qin
Swarm optimization and lens design

Q. Ramasse, R. Brydson
The SuperSTEM laboratory

B. Rieger, A.J. Koster
Image formation in cryo-electron microscopy

P. Rocca, M. Donelli
Imaging of dielectric objects

J. Rodenburg
Lensless imaging

J. Rouse, H.-n. Liu, E. Munro
The role of differential algebra in electron optics

J. Sánchez
Fisher vector encoding for the classification of natural images

P. Santi
Light sheet fluorescence microscopy

R. Shimizu, T. Ikuta, Y. Takai
Defocus image modulation processing in real time

T. Soma
Focus-deflection systems and their applications

J. Valdés
Recent developments concerning the Système International (SI)

CHAPTER ONE

New Physical Principle for Interference of Light and Material Particles

Román Castañeda*[,1], Giorgio Matteucci†
*Physics School, Universidad Nacional de Colombia Sede Medellín, Medellín, Colombia
†University of Bologna, Bologna, Italy
[1]Corresponding author: e-mail address: rcastane@unal.edu.co

Contents

1. INTRODUCTION

The wave properties of light and matter are usually discussed by considering diffraction and interference effects. Interference patterns of light or of material particles are explained using, respectively, propagating electromagnetic fields (Born & Wolf, 1993), or complex probability waves (mathematical wave functions without a direct physical meaning) (Feynman, Leighton, & Sands, 1965).

Although analogous, natural interference effects, observed with light or material particles, are described with the superposition principle, these effects are not explained with the same cause as required by the Newton's second rule of Reasoning and Philosophy. According to Newton, analogous interference fringes of waves or of material particles should be produced by the same cause and possibly be described with a unified model. In addition, a success along this line of thought could result in the removal of the contestable features regarding the physical meaning of the probability wave function.

Advances in Imaging and Electron Physics, Volume 204
ISSN 1076-5670
https://doi.org/10.1016/bs.aiep.2017.09.001

With this in mind, here we propose a generalized interference principle of classical waves or of massive particles on the basis of the classical spatial correlation theory (Mandel & Wolf, 1995). The principle takes into consideration individual interactions between *real point emitters*, that represent a wave disturbance or particles passing through one or more apertures, and *virtual point emitters* (modulating sources of energy) determined by the setup configuration (Castañeda, 2014; Castañeda, Matteucci, & Capelli, 2016a). With this model, the wave superposition principle together with the controversial hypotheses regarding the interpretation of the wave function, adopted to account for the behavior of material particles, are not necessary requirements to describe interference (Castañeda, Matteucci, & Capelli, 2016b). In order to synthesize the principle, important modifications of the conventional interpretation of the mathematical model are introduced. In section 2, a novel theoretical model regarding the modal expansion of the spatial correlation is interpreted as a geometrical condition of the space delimited by the setup configuration. In section 3, a new general law of interference is proposed for waves and particles moving in field free regions and *a geometric potential* is introduced as physical condition for interference (Castañeda, 2017b). The main differences between the general law of interference and the conventional one are detailed in section 4. In section 5, the *spectrum of classes of point emitters* (Castañeda & Muñoz, 2016) is presented as a theoretical mean to construct complete maps of interactions, between real and virtual point emitters, to describe any interference experiment. Original preliminary considerations, regarding the accurate prediction of diffraction envelopes and the influence of the uncertainty in position of particles going through the slits of a grating are reported in section 6. Profiles of experimental interference patterns with single molecules are presented to demonstrate the validity of the present theory. In the Conclusions, the relevant results are recalled concerning the role played by the spatial geometrical properties of an interferometer in the formation of interference patterns and the connection of the uncertainty principle with the diffraction envelope.

2. A NOVEL THEORETICAL MODEL

The most sophisticated theory that explains interference of light is based on Maxwell's wave equations. In this context, the time-independent (or static) part of the wave equation is a Helmholtz equation, whose eigen-functions have been interpreted as the wave disturbance in each point of space. Thus, interference of light is accurately accounted for as the superposition of the eigen-functions of the Helmholtz's equation in the framework of the Maxwell's electrodynamics (Born & Wolf, 1993).

Subsequently, massive particle interference was theoretically described by solving the time-independent part of the Schrödinger equation in a field-free region, which is also a Helmholtz equation, with similar boundary conditions used for wave interference (Feynman et al., 1965). Although the theoretical predictions have been validated by experimental results (Arndt et al., 1999; Bach, Pope, Liou, & Batelaan, 2013; Frabboni et al., 2012; Juffmann et al., 2009; Juffmann et al., 2012; Matteucci, 2013; Matteucci et al., 2013; Nairz, Arndt, & Zeilinger, 2003; Zeilinger, Gaehler, Shull, Treimer, & Mampe, 1988), the various attempts to attribute a physical meaning to the eigen-functions of this Helmholtz equation have not been successful.

Here, we will show that a new perspective of the two-point correlation of the Helmholtz eigen-functions provides a unified careful explanation of wave and particle interference without resorting to the superposition principle. For this purpose, let us start by considering a canonical configuration of interference experiments consisting in a two-stage setup as sketched in Fig. 1. The SM-stage is the volume delimited by the source S and the mask M planes placed at a distance z' to each other. At the M-plane an interferometry device, for instance a grating, is usually inserted. The SM-stage is called the *preparation stage* because it provides the waves or particles with a specific energy distribution and determines the two-point correlation condition at the M-plane needed to obtain the final interference pattern in the MD-stage. The second stage MD is confined between M and the detector plane D, placed at a distance z, where a conventional squared modulus detector is located. The MD-stage is called the *realization stage*, because it determines the energy distribution, at the D-plane, that represents the physical observable (wave or particle interference pattern).

The center and difference coordinates, $(\mathbf{r}'_A, \mathbf{r}'_D)$ for the S-plane, $(\boldsymbol{\xi}_A, \boldsymbol{\xi}_D)$ for the M-plane and $(\mathbf{r}_A, \mathbf{r}_D)$ for the D-plane, allow determining univocally pairs of points on the respective plane, with separation vector given by the coordinate with suffix D. They are equidistant to the point specified by the coordinate with suffix A. So, their positions are $\mathbf{r}'_\pm = \mathbf{r}'_A \pm \mathbf{r}'_D/2$, $\boldsymbol{\xi}_\pm = \boldsymbol{\xi}_A \pm \boldsymbol{\xi}_D/2$, and $\mathbf{r}_\pm = \mathbf{r}_A \pm \mathbf{r}_D/2$ respectively.

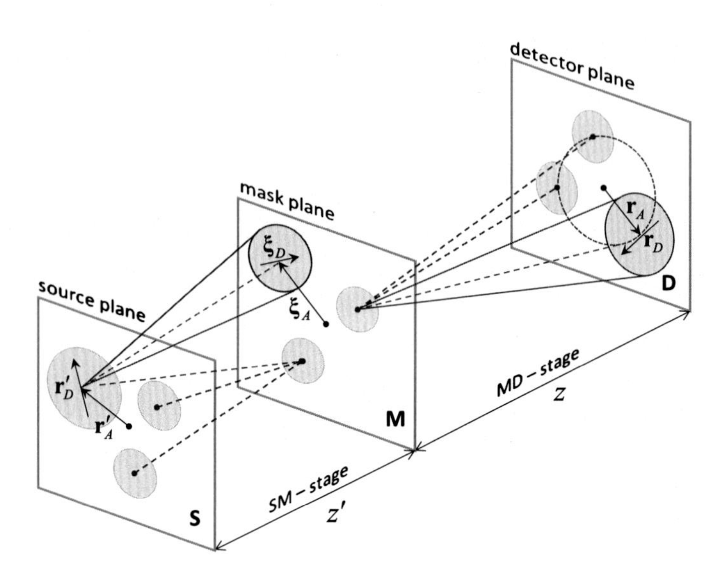

Figure 1 Coordinate system used in the two-stage interference setup. Shadowed circles on each plane represent the structured supports of correlation. Cones depicted with solid lines represent correlation cones.

In previous works (Castañeda et al., 2016a, 2016b) we have demonstrated that interference of waves as well as of single material particles is accurately described by two-point correlation functions. Specifically, the physical observable, i.e. the energy distribution of the interference pattern recorded by a conventional squared modulus detector is calculated with the two-point correlation of the eigen-functions $\psi(\mathbf{r}_j)$ of two Helmholtz equations coupled by their eigen-values (Mandel & Wolf, 1995), which have the form $\nabla_j^2\psi(\mathbf{r}_j) = -k^2\psi(\mathbf{r}_j)$, with $j = 1, 2$. The Laplacian operator ∇_j^2 depends on the $\mathbf{r}_j$ coordinates and $k = \omega/c$ for a wave of frequency ω and propagation speed c, while $k = p/\hbar$ with $\hbar = h/2\pi$ and h the Planck constant, for a particle of momentum p.

The solution of the coupled Helmholtz equations, by means of the Green's function method (Mandel & Wolf, 1995) under the specific boundary conditions of each setup stage depicted in Fig. 1, gives the two-point correlation of the eigen-functions at the exit plane of the corresponding stage, $W_M(\boldsymbol{\xi}_+, \boldsymbol{\xi}_-) = \langle\psi(\boldsymbol{\xi}_+)\psi^*(\boldsymbol{\xi}_-)\rangle$ and $W_D(\mathbf{r}_+, \mathbf{r}_-) = \langle\psi(\mathbf{r}_+)\psi^*(\mathbf{r}_-)\rangle$. The symbol $\langle\rangle$ and the asterisk denote ensemble average and complex conjugated respectively. Thus, the energy distribution of the interference pattern recorded by the detector at the D-plane is determined by evaluating the two-point correlation at such plane for $\mathbf{r}_D = 0$, i.e. $S_D(\mathbf{r}_A) = W_D(\mathbf{r}_A, \mathbf{r}_A) = \langle|\psi(\mathbf{r}_A)|^2\rangle$. The correlation on each plane is related to regions of pairs of points called *structured supports of correlation*, (in optics, structured supports of spatial coherence) (Castañeda, 2014; Castañeda et al., 2016b), which are centered at the position labeled by the coordinate with suffix A, Fig. 1. It is important to remark that the structured supports are not equivalent to the conventional coherence areas, spots or patches, used in optics to characterize correlation (Born & Wolf, 1993). Indeed, each structured support, which is identified by the position of its center, labeled by the coordinate suffixed A, gathers pairs of points symmetrically located with respect to the center. Only these pairs contribute to the correlation. It means that the central point and the pairs which are not symmetrically placed with respect to the support center cannot be accounted for the two-point correlation over the considered structured support.

The preparation of the experimental conditions which must be satisfied in the SM-stage consists in calculating the two-point correlation at the M-plane, once the two-point correlation at the S-plane is known, Fig. 1. Thereafter, the experiment realization in the MD-stage defines how energy distributes on the interference pattern at the D-plane. This process is described starting from the two-point correlation at the M-plane followed by calculation of the two-point correlation at the D-plane.

Let us start to describe the two-point correlation distribution in the preparation stage SM. The two-point correlation at the M-plane is given by

$$W_M(\boldsymbol{\xi}_+, \boldsymbol{\xi}_-) = \int_S d^2 r'_A \, \mathbf{W}_{SM}\big(\mathbf{r}'_A; \boldsymbol{\xi}_+, \boldsymbol{\xi}_-\big), \tag{1a}$$

with the modal expansion

$$\mathbf{W}_{SM}\big(\mathbf{r}'_A; \boldsymbol{\xi}_+, \boldsymbol{\xi}_-\big) = \int_S d^2 r'_D \, W_S\big(\mathbf{r}'_+, \mathbf{r}'_-\big) \Phi_{SM}\big(\mathbf{r}'_+, \mathbf{r}'_-; \boldsymbol{\xi}_+, \boldsymbol{\xi}_-; k, z'\big). \tag{1b}$$

The expansion coefficient $W_S(\mathbf{r}'_+, \mathbf{r}'_-)$ denotes the two-point correlation at the S-plane, and its kernel is defined by the scalar, deterministic, and non-paraxial modes, determined by the stage configuration (Castañeda, 2014; Castañeda et al., 2016a),

$$\Phi_{SM}(\mathbf{r}'_+, \mathbf{r}'_-; \boldsymbol{\xi}_+, \boldsymbol{\xi}_-; k, z') = \left(\frac{k}{4\pi}\right)^2 t_S(\mathbf{r}'_+) t_S^*(\mathbf{r}'_-)$$
$$\times \left(\frac{z' + |\mathbf{z}' + \boldsymbol{\xi}_A - \mathbf{r}'_A + (\boldsymbol{\xi}_D - \mathbf{r}'_D)/2|}{|\mathbf{z}' + \boldsymbol{\xi}_A - \mathbf{r}'_A + (\boldsymbol{\xi}_D - \mathbf{r}'_D)/2|^2}\right)$$
$$\times \left(\frac{z' + |\mathbf{z}' + \boldsymbol{\xi}_A - \mathbf{r}'_A - (\boldsymbol{\xi}_D - \mathbf{r}'_D)/2|}{|\mathbf{z}' + \boldsymbol{\xi}_A - \mathbf{r}'_A - (\boldsymbol{\xi}_D - \mathbf{r}'_D)/2|^2}\right)$$
$$\times \exp\big(ik|\mathbf{z}' + \boldsymbol{\xi}_A - \mathbf{r}'_A + (\boldsymbol{\xi}_D - \mathbf{r}'_D)/2|$$
$$- ik|\mathbf{z}' + \boldsymbol{\xi}_A - \mathbf{r}'_A - (\boldsymbol{\xi}_D - \mathbf{r}'_D)/2|\big), \qquad (1c)$$

with

(i) $k = 2\pi/\lambda$ and λ the space scale metric of the setup, defined as the length along which the propagator argument of the modes evolves in 2π, in accordance to the boundary conditions (Castañeda, 2014; Castañeda et al., 2016a, 2016b). It means that only waves of frequency $\omega = ck = 2\pi c/\lambda$, photons of energy $E = \hbar\omega = \hbar ck = hc/\lambda$ or particles of momentum $p = \hbar k = h/\lambda$ can propagate through the modes with parameter $k = 2\pi/\lambda$. It must be noted that the interpretation of λ as a setup parameter that couples waves and particles to the setup through their physical attributes (frequency, energy, or momentum) allows to remove the requirement of wave–particle dualism to explain interference, as we will see presently and,

(ii) $t_S(\mathbf{r}'_\pm) = |t_S(\mathbf{r}'_\pm)| \exp[i\phi_S(\mathbf{r}'_\pm)]$ the complex transmission of the effective source of waves or particles at the S-plane. It delimits the region in which waves or particles are emitted at the S-plane.

The modes in Eq. (1c) are defined in the whole volume of the SM-stage independently from the specific form of $W_S(\mathbf{r}'_+, \mathbf{r}'_-)$. So, the expansion kernel shapes the geometry for the propagation in this stage for any type of effective source. Furthermore, $W_S(\mathbf{r}'_+, \mathbf{r}'_-) = \sqrt{S_S(\mathbf{r}'_+)}\sqrt{S_S(\mathbf{r}'_-)}\mu_S(\mathbf{r}'_+, \mathbf{r}'_-)$, where $S_S(\mathbf{r}'_\pm)$ represents the average energy of the local emissions of the effective source, and $\mu_S(\mathbf{r}'_+, \mathbf{r}'_-) = |\mu_S(\mathbf{r}'_+, \mathbf{r}'_-)| \exp[i\alpha_S(\mathbf{r}'_+, \mathbf{r}'_-)]$ is the correlation degree at the S-plane, with $0 \leq |\mu_S(\mathbf{r}'_+, \mathbf{r}'_-)| \leq 1$, $\mu_S(\mathbf{r}'_A, \mathbf{r}'_A) = 1$, $\alpha_S(\mathbf{r}'_+, \mathbf{r}'_-) = \arg[W_S(\mathbf{r}'_+, \mathbf{r}'_-)]$, and $\mu_S(\mathbf{r}'_+, \mathbf{r}'_-) = \mu_S^*(\mathbf{r}'_-, \mathbf{r}'_+)$.

The above definition of $W_S(\mathbf{r}'_+, \mathbf{r}'_-)$ is conventionally established from its achievement of the Schwartz inequality (Mandel & Wolf, 1995). Alternatively, it can also be obtained by expressing $\psi(\mathbf{r}'_\pm) = \sqrt{S_S(\mathbf{r}'_\pm)}\psi_N(\mathbf{r}'_\pm)$ where $\psi_N(\mathbf{r}'_\pm)$ is the normalized eigen-function of the Helmholtz equations at the S-plane, which is a geometrical function. Therefore, the two-point correlation at the S-plane becomes $W_S(\mathbf{r}'_+, \mathbf{r}'_-) = \langle \psi(\mathbf{r}'_+)\psi^*(\mathbf{r}'_-)\rangle = \sqrt{S_S(\mathbf{r}'_+)}\sqrt{S_S(\mathbf{r}'_-)}\langle \psi_N(\mathbf{r}'_+)\psi_N^*(\mathbf{r}'_-)\rangle$, thus being $\mu_S(\mathbf{r}'_+, \mathbf{r}'_-) = \langle \psi_N(\mathbf{r}'_+)\psi_N^*(\mathbf{r}'_-)\rangle$

the correlation degree at the S-plane. The advantage of this alternative deduction over the conventional one is the geometrical meaning that it gives to the correlation degree, in connection with the emission events of the effective source at the S-plane. In the conventional deduction, the correlation degree is meaningless and introduced *ad hoc*.

The integration domain of Eq. (1b) is the structured support of correlation centered at $\mathbf{r}'_A$, which is also the correlation degree support, i.e. the region in which $\mu_S(\mathbf{r}'_+, \mathbf{r}'_-)$ is different from zero. The correlation degree behaves, therefore, like a modal filter that weights the contribution of the non-paraxial modes in the expansion kernel. Indeed, the expansion model for $\mu_S(\mathbf{r}'_+, \mathbf{r}'_-) = 1$ defines the maximum set of modes, in the stage configuration, for any pair of points on the effective source area (i.e. spatially correlated emission process).

A subset of modes accounts for the kernel if the structured support size is smaller than the effective source size. The extension of the mode subset becomes minimum when $\mu_S(\mathbf{r}'_+, \mathbf{r}'_-) = 0$ for any $\mathbf{r}'_D \neq 0$. This is the case of a spatial uncorrelated emission process. It is worth noting that the physic and statistical properties of Eq. (1a) are attributes of the emission process of the effective source represented by $W_S(\mathbf{r}'_+, \mathbf{r}'_-)$. Nevertheless, such attributes have no influence on the geometry of the modes which varies only if the stage configuration is modified.

In Eq. (1b), $\mathbf{W}_{SM}(\mathbf{r}'_A; \boldsymbol{\xi}_+, \boldsymbol{\xi}_-)$ has energy units and, according to its arguments, describes the contribution from each given point $\mathbf{r}'_A$, of the effective source, onto the structured support of correlation centered at any given point $\boldsymbol{\xi}_A$ on the M-plane. In other words, $\mathbf{W}_{SM}(\mathbf{r}'_A; \boldsymbol{\xi}_+, \boldsymbol{\xi}_-)$ determines a cone with vertex on $\mathbf{r}'_A$ and basis on the structured support centered at $\boldsymbol{\xi}_A$ whose geometry, in the whole SM-stage, is determined by the expansion kernel (Fig. 1).

The experiment preparation with uncorrelated emission process of waves or particles is of paramount importance. In this case, Eq. (1b) yields

$$\mathbf{W}_{SM}\big(\mathbf{r}'_A; \boldsymbol{\xi}_+, \boldsymbol{\xi}_-\big) = S_S\big(\mathbf{r}'_A\big)\Phi_{SM}\big(\mathbf{r}'_A; \boldsymbol{\xi}_+, \boldsymbol{\xi}_-; k, z'\big), \tag{2a}$$

with $S_S(\mathbf{r}'_A) = W_S(\mathbf{r}'_A, \mathbf{r}'_A)$, so that Eq. (1a) becomes

$$W_M(\boldsymbol{\xi}_+, \boldsymbol{\xi}_-) = \int_S d^2r'_A\, S_S\big(\mathbf{r}'_A\big)\Phi_{SM}\big(\mathbf{r}'_A; \boldsymbol{\xi}_+, \boldsymbol{\xi}_-; k, z'\big). \tag{2b}$$

It means that the two-point correlation at the M-plane takes non-null values despite the fact that the two-point correlation at the S-plane nullifies for any pair of points. In other words, the gain of spatial correlation at the M-plane is caused by the geometry of the modal kernel in the SM-stage. This outstanding outcome, which is due to the cone geometry of the modal expansion in Eq. (2b), results from the overlapping of the cones defined by Eq. (2a). Eq. (2b) is known in optics as the non-paraxial Van Cittert–Zernike theorem (Born & Wolf, 1993; Castañeda, 2017b), and describes the preparation conditions for single particle interference experiments.

The conventional explanation of this theorem in optics attributes the gain of spatial correlation to physic and statistical properties of the propagating waves (Born & Wolf, 1993; Mandel & Wolf, 1995). However, it is not applicable in the same sense to particle interference. Indeed, although the two-point correlation of the quantum wave function can be calculated, this wave function has no specific physical meaning. In contrast, the geometrical explanation of Eq. (2b) is independent of attributes of waves or particles moving in the SM-stage and, as we will demonstrate, it has a high prediction accuracy of experimental results. In addition, Eq. (1b) can be expressed as

$$\mathbf{W}_{SM}(\mathbf{r}'_A;\boldsymbol{\xi}_+,\boldsymbol{\xi}_-) = S_S(\mathbf{r}'_A)\Phi_{SM}(\mathbf{r}'_A;\boldsymbol{\xi}_+,\boldsymbol{\xi}_-;k,z') + \int_{\mathbf{r}'_D\neq 0} d^2r'_D\, W_S(\mathbf{r}'_+,\mathbf{r}'_-)\Phi_{SM}(\mathbf{r}'_+,\mathbf{r}'_-;\boldsymbol{\xi}_+,\boldsymbol{\xi}_-;k,z').$$

The first term, represented by Eq. (2a), describes an uncorrelated emission process while the second one refers to partial correlated emission events. It means that the shape and size of the correlation increase when more modes are included in the modal expansion. Consequently, the respective resulting cones in the SM-stage have different geometries determined by Eq. (1a). The energy distribution at the M-plane is given by evaluating Eq. (1a) for $\boldsymbol{\xi}_D = 0$ i.e. $S_M(\boldsymbol{\xi}_A) = W_M(\boldsymbol{\xi}_A, \boldsymbol{\xi}_A)$.

A completely similar analysis is applicable to the realization stage MD. We will show how the energy distribution of waves or particles on the interference pattern, formed at the D-plane, can be calculated. The physical observable, recorded by a squared modulus detector, is given by the modal expansion

$$S_D(\mathbf{r}_A) = \int_M d^2\xi_A\, \mathbf{W}_{MD}(\boldsymbol{\xi}_A;\mathbf{r}_A), \tag{3a}$$

with

$$\mathbf{W}_{MD}(\boldsymbol{\xi}_A; \mathbf{r}_A) = \int_M d^2\xi_D \, W_M(\boldsymbol{\xi}_+, \boldsymbol{\xi}_-)\Phi_{MD}(\boldsymbol{\xi}_+, \boldsymbol{\xi}_-; \mathbf{r}_A; k, z), \tag{3b}$$

where

$$\begin{aligned}\Phi_{MD}(\boldsymbol{\xi}_+, \boldsymbol{\xi}_-; \mathbf{r}_A; k, z) = {} & \left(\frac{k}{4\pi}\right)^2 t_M(\boldsymbol{\xi}_+)t_M^*(\boldsymbol{\xi}_-) \\ & \times \left(\frac{z + |\mathbf{z} + \mathbf{r}_A - \boldsymbol{\xi}_A - \boldsymbol{\xi}_D/2|}{|\mathbf{z} + \mathbf{r}_A - \boldsymbol{\xi}_A - \boldsymbol{\xi}_D/2|^2}\right) \\ & \times \left(\frac{z + |\mathbf{z} + \mathbf{r}_A - \boldsymbol{\xi}_A + \boldsymbol{\xi}_D/2|}{|\mathbf{z} + \mathbf{r}_A - \boldsymbol{\xi}_A + \boldsymbol{\xi}_D/2|^2}\right) \\ & \times \exp\big(ik|\mathbf{z} + \mathbf{r}_A - \boldsymbol{\xi}_A - \boldsymbol{\xi}_D/2| \\ & \qquad - ik|\mathbf{z} + \mathbf{r}_A - \boldsymbol{\xi}_A + \boldsymbol{\xi}_D/2|\big)\end{aligned} \tag{3c}$$

is the modal kernel determined by the MD-stage configuration, with $t_M(\boldsymbol{\xi}_\pm) = |t_M(\boldsymbol{\xi}_\pm)|\exp[i\phi_M(\boldsymbol{\xi}_\pm)]$ the complex transmission of the interference mask attached at the M-plane, and $\Phi_{MD}(\boldsymbol{\xi}_+, \boldsymbol{\xi}_-; \mathbf{r}_A; k, z) = \Phi^*_{MD}(\boldsymbol{\xi}_-, \boldsymbol{\xi}_+; \mathbf{r}_A; k, z)$ describes the scalar, deterministic, and non-paraxial modes defined in the volume of the MD-stage, independently from the specific form of $W_M(\boldsymbol{\xi}_+, \boldsymbol{\xi}_-)$. They shape the geometry of the volume which changes only if the stage configuration changes. The modal expansion in Eq. (3b) is calculated over the structured support of correlation centered at each $\boldsymbol{\xi}_A$ on the interference mask at the M-plane. The coefficient of $W_M(\boldsymbol{\xi}_+, \boldsymbol{\xi}_-)$ is determined by Eq. (2b) and confers its statistical properties for the buildup of an interference pattern. In addition, $W_M(\boldsymbol{\xi}_+, \boldsymbol{\xi}_-) = \sqrt{S_M(\boldsymbol{\xi}_+)}\sqrt{S_M(\boldsymbol{\xi}_-)}\mu_M(\boldsymbol{\xi}_+, \boldsymbol{\xi}_-)$ with $S_M(\boldsymbol{\xi}_\pm) = W_M(\boldsymbol{\xi}_\pm, \boldsymbol{\xi}_\pm)$ the average energy emerging from each point of the mask, and $\mu_M(\boldsymbol{\xi}_+, \boldsymbol{\xi}_-) = |\mu_M(\boldsymbol{\xi}_+, \boldsymbol{\xi}_-)|\exp[i\alpha_M(\boldsymbol{\xi}_+, \boldsymbol{\xi}_-)]$ the correlation degree at the M-plane, which has similar mathematical properties of the correlation degree at the S-plane.

Therefore, the modal expansion for $\mathbf{W}_{MD}(\boldsymbol{\xi}_A; \mathbf{r}_A)$ in Eq. (3b) has the following important features:

(i) The two-point correlation at the M-plane (and specifically the degree of correlation) behaves as a modal filter that selects the modes for the MD-stage.

(ii) It has energy units.

(iii) Its arguments, defined in the volume of the MD-stage, describe the contribution from each point $\boldsymbol{\xi}_A$ in the mask onto any point $\mathbf{r}_A$ of the interference pattern. Therefore, it determines a cone with vertex on $\boldsymbol{\xi}_A$ and base including the points $\mathbf{r}_A$ which constitute the interference pattern at the D-plane (dotted line cone in the MD-stage, Fig. 1).

Finally, Eq. (3a) describes the energy of the interference pattern distributed over the cross-section of the cone which results from the overlapping, at the D-plane, of the $\mathbf{W}_{MD}(\boldsymbol{\xi}_A; \mathbf{r}_A)$ cones.

3. GENERAL LAW OF INTERFERENCE

The considerations developed in the previous section are applied to generalize interference of waves and massive particles moving in field free regions of the Young's setup depicted in Fig. 2. Firstly, the configuration of the two-point correlation in the SM preparation stage is calculated. The openings in the mask at the S and M-planes are pinholes, so that

$$t_S(\mathbf{r}'_+)t_S(\mathbf{r}'_-) = \delta(\mathbf{r}'_A)\delta(\mathbf{r}'_D), \tag{4a}$$

with $\delta(\bullet)$ the Dirac's delta, and

$$\begin{aligned} t_M(\boldsymbol{\xi}_+)t_M(\boldsymbol{\xi}_-) = {} & \left[\delta(\boldsymbol{\xi}_A + \mathbf{b}/2) + \delta(\boldsymbol{\xi}_A - \mathbf{b}/2)\right]\delta(\boldsymbol{\xi}_D) \\ & + \left[\delta(\boldsymbol{\xi}_D + \mathbf{b}) + \delta(\boldsymbol{\xi}_D - \mathbf{b})\right]\delta(\boldsymbol{\xi}_A). \end{aligned} \tag{4b}$$

Eqs. (1b) and (4a) yield

$$\begin{aligned} \mathbf{W}_{SM}(\mathbf{r}'_A; \boldsymbol{\xi}_+, \boldsymbol{\xi}_-) = {} & \left(\frac{k}{4\pi}\right)^2 S_S(\mathbf{r}'_A)\delta(\mathbf{r}'_A)\left(\frac{z' + |\mathbf{z}' + \boldsymbol{\xi}_A - \mathbf{r}'_A + \boldsymbol{\xi}_D/2|}{|\mathbf{z}' + \boldsymbol{\xi}_A - \mathbf{r}'_A + \boldsymbol{\xi}_D/2|^2}\right) \\ & \times \left(\frac{z' + |\mathbf{z}' + \boldsymbol{\xi}_A - \mathbf{r}'_A - \boldsymbol{\xi}_D/2|}{|\mathbf{z}' + \boldsymbol{\xi}_A - \mathbf{r}'_A - \boldsymbol{\xi}_D/2|^2}\right) \\ & \times \exp\left(ik|\mathbf{z}' + \boldsymbol{\xi}_A - \mathbf{r}'_A + \boldsymbol{\xi}_D/2| - ik|\mathbf{z}' + \boldsymbol{\xi}_A - \mathbf{r}'_A - \boldsymbol{\xi}_D/2|\right). \end{aligned} \tag{5}$$

The geometric factor of Eq. (5) describes correlation cones with vertex at the pinhole on $\mathbf{r}'_A = 0$, at the S-plane, and bases on the structured support of correlation centered at any $\boldsymbol{\xi}_A$ on the M-plane (Fig. 2). The axes of the correlation cones determine the cones described by the geometric factor

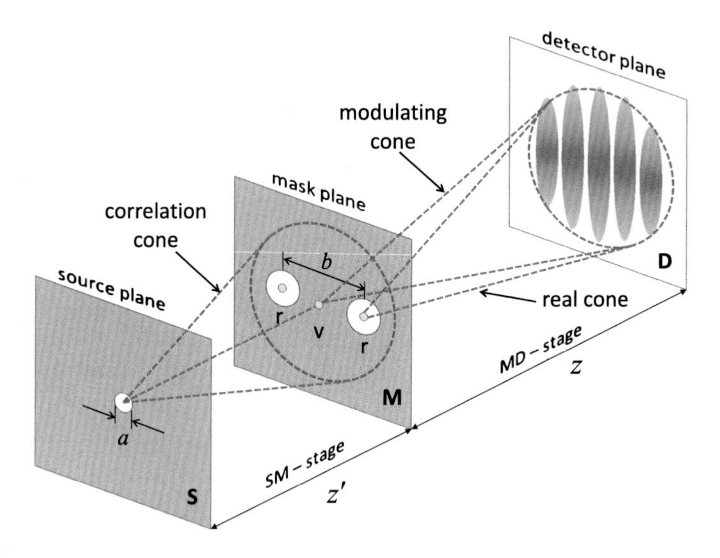

Figure 2 Conceptual sketch of the Young interferometer. Real and virtual point emitters are labeled as r and v respectively. The physical meaning of the correlation cone, of the real cone, and of the modulating cone is explained in the text.

of the expression

$$\mathbf{W}_{SM}(\mathbf{r}'_A; \boldsymbol{\xi}_A) = \left(\frac{k}{4\pi}\right)^2 S(\mathbf{r}'_A)\delta(\mathbf{r}'_A)\left(\frac{z' + |\mathbf{z}' + \boldsymbol{\xi}_A - \mathbf{r}'_A|}{|\mathbf{z}' + \boldsymbol{\xi}_A - \mathbf{r}'_A|^2}\right)^2, \tag{6}$$

whose bases, at the M-plane, delimit the regions of points $\boldsymbol{\xi}_A$ where the energy emitted by the effective source is localized. Examples of correlation and energy cones for the Young setup of Fig. 2 are shown in Fig. 3, in case of light waves and single massive particles, respectively. Distinctive features of these cones, independent of the value of λ, are, (i) their shape invariant Lorentzian cross-sections for $z' \geq 0$ and, (ii) their characteristic aperture angles with respect to the axis.

The Lorentzian profile, which is a consequence of the delta like geometry assumed for the pinhole at the S-plane, Fig. 2, decays monotonically and symmetrically with respect to the cone axis. It is reasonable to choice $\sigma/2$ (the radius of the structured support of correlation) and $\varepsilon/2$ (the radius of the energy cone base, at given z') as the values of $\boldsymbol{\xi}_D$ and $\boldsymbol{\xi}_A$ respectively, for which the height of the profiles decays of 90%. Therefore, the effective aperture angles are ~76° for the correlation cone and ~63° for the energy cone for both waves and massive particles. It is remarkable that the finite

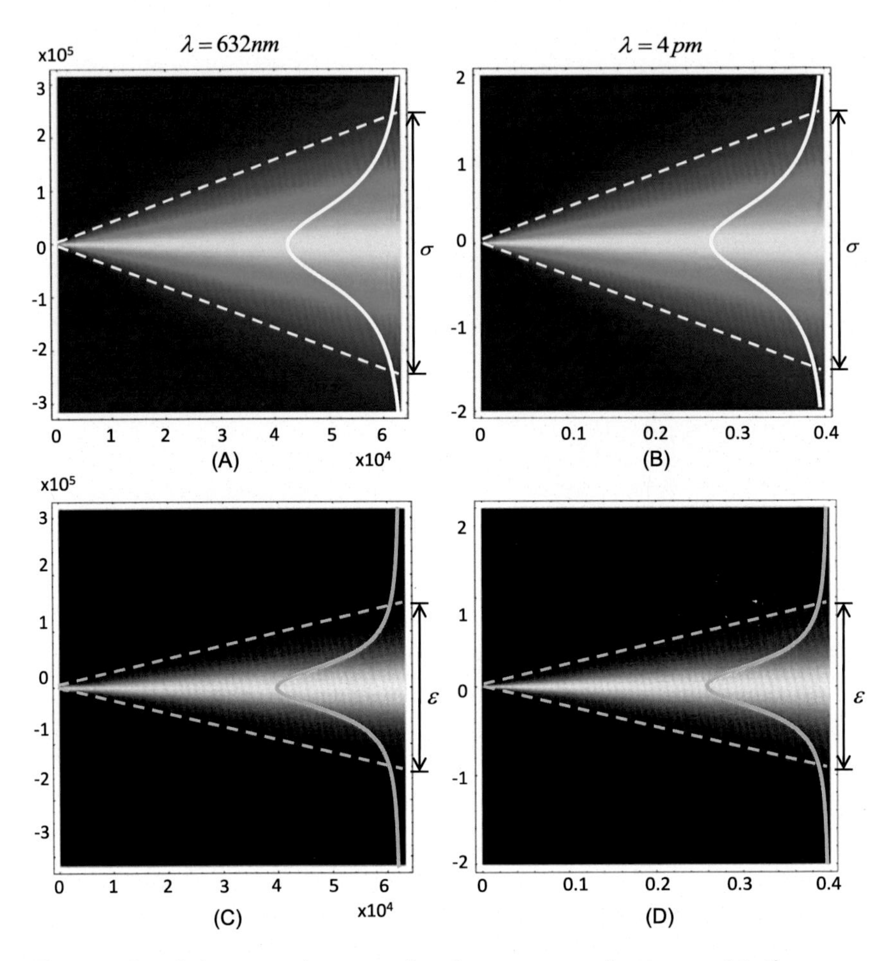

Figure 3 Correlation cones (upper row) and energy cones (bottom row) in the preparation stage of a Young interferometer for light waves (left column) and single massive particles (right column). In all graphs the axes are in μm, the horizontal axis is z' and the cone vertices are at $r'_A = 0$. The vertical axes are r'_D at $z' = 0$ and ξ_D otherwise for the correlation cones (A) and (B), and r'_A at $z' = 0$ and ξ_A otherwise for the energy cones in (C) and (D). The cross sections of all the cones at any z' is Lorentzian-like and their aperture angles with respect to the cone axes are $\sim 76°$ for the correlation cones and $\sim 63°$ for the energy cones. σ and ε denote respectively the diameter of the structured support at the bases of the correlation cones and the diameter of the energy cone bases. Graphs were enhanced for presentation purposes.

aperture angle of these non-paraxial cones is quite different from the value obtained with the standard optical description being it limited to the paraxial regime. Indeed, in far-field and paraxial conditions, a point-like source provides a coherent and uniform illumination over an arbitrary large region (Born & Wolf, 1993).

The aperture angles of the correlation cones are controlled by the size and shape of $t_S(\mathbf{r}'_+)t^*_S(\mathbf{r}'_-)$. The maximum angle is obtained with delta-like transmission functions $t_S(\mathbf{r}'_+)t^*_S(\mathbf{r}'_-)$ in Eq. (4a). For increasing sizes of $t_S(\mathbf{r}'_+)t^*_S(\mathbf{r}'_-)$ the angles decrease. So, Eq. (1a) gives the two-point correlation at the mask plane,

$$W_M(\boldsymbol{\xi}_+,\boldsymbol{\xi}_-) = \left(\frac{k}{4\pi}\right)^2 S_S(0) \\ \times \left(\frac{z' + |\mathbf{z}' + \boldsymbol{\xi}_A + \boldsymbol{\xi}_D/2|}{|\mathbf{z}' + \boldsymbol{\xi}_A + \boldsymbol{\xi}_D/2|^2}\right)\left(\frac{z' + |\mathbf{z}' + \boldsymbol{\xi}_A - \boldsymbol{\xi}_D/2|}{|\mathbf{z}' + \boldsymbol{\xi}_A - \boldsymbol{\xi}_D/2|^2}\right) \\ \times \exp\left(ik\left|\mathbf{z}' + \boldsymbol{\xi}_A + \boldsymbol{\xi}_D/2\right| - ik\left|\mathbf{z}' + \boldsymbol{\xi}_A - \boldsymbol{\xi}_D/2\right|\right). \tag{7}$$

Eq. (7) details the parameters of the preparation conditions. It accounts for, (i) the structured support to which the pinhole pair belongs and, (ii) the energy at each pinhole once the effective source is active.

The structured support is determined by Eq. (7) for $\boldsymbol{\xi}_A = 0$ and the specific correlation of the pinholes by taking $\boldsymbol{\xi}_D = \mathbf{b}$, that is

$$W_M(\mathbf{b}) = \left(\frac{k}{4\pi}\right)^2 S_S(0)\left(\frac{z' + |\mathbf{z}' + \mathbf{b}/2|}{|\mathbf{z}' + \mathbf{b}/2|^2}\right)\left(\frac{z' + |\mathbf{z}' - \mathbf{b}/2|}{|\mathbf{z}' - \mathbf{b}/2|^2}\right) \\ \times \exp\left(ik\left|\mathbf{z}' + \mathbf{b}/2\right| - ik\left|\mathbf{z}' - \mathbf{b}/2\right|\right). \tag{8a}$$

The energy at each pinhole can be determined by evaluating Eq. (7) for $\boldsymbol{\xi}_A = \pm\mathbf{b}/2$ and $\boldsymbol{\xi}_D = 0$, i.e.

$$S_M(\pm\mathbf{b}/2) = \left(\frac{k}{4\pi}\right)^2 S_S(0)\left(\frac{z' + |\mathbf{z}' \pm \mathbf{b}/2|}{|\mathbf{z}' \pm \mathbf{b}/2|^2}\right)^2. \tag{8b}$$

It takes on the same value in both pinholes at a fixed distance z' from the S-plane. Thus, Eqs. (8) represent the Young's experiment preparation. Because of the symmetry of the Lorentzian cross-sections of the cones with respect to the z-axis, $|\mathbf{b}| < \varepsilon$ should be fulfilled in order to assure that energy from the source affects both pinholes, even in case of independent single emission events. The accomplishment of this condition implies $|\mathbf{b}| < \sigma$ because the aperture angle of the energy cone is narrower than the aperture angle of the correlation cone.

The realization conditions of the experiment are described by Eqs. (3b) and (4b). They allow to determine the energy cone in the MD-stage as $\mathbf{W}_{MD}(\boldsymbol{\xi}_A;\mathbf{r}_A)=\mathbf{W}_{MD}^{(+)}(\boldsymbol{\xi}_A;\mathbf{r}_A)+\mathbf{W}_{MD}^{(-)}(\boldsymbol{\xi}_A;\mathbf{r}_A)$ with

$$
\begin{aligned}
\mathbf{W}_{MD}^{(\pm)}(\boldsymbol{\xi}_A;\mathbf{r}_A) &= \left(\frac{k}{4\pi}\right)^2 S_M(\boldsymbol{\xi}_A)\left(\frac{z+|\mathbf{z}+\mathbf{r}_A-\boldsymbol{\xi}_A|}{|\mathbf{z}+\mathbf{r}_A-\boldsymbol{\xi}_A|^2}\right)^2 \delta(\boldsymbol{\xi}_A\mp\mathbf{b}/2)\\
&+\left(\frac{k}{4\pi}\right)^2 |\mu_M(\boldsymbol{\xi}_A+\mathbf{b}/2,\boldsymbol{\xi}_A-\mathbf{b}/2)|\sqrt{S_M(\boldsymbol{\xi}_A+\mathbf{b}/2)}\\
&\times\sqrt{S_M(\boldsymbol{\xi}_A-\mathbf{b}/2)}\\
&\times\left(\frac{z+|\mathbf{z}+\mathbf{r}_A-\boldsymbol{\xi}_A-\mathbf{b}/2|}{|\mathbf{z}+\mathbf{r}_A-\boldsymbol{\xi}_A-\mathbf{b}/2|^2}\right)\left(\frac{z+|\mathbf{z}+\mathbf{r}_A-\boldsymbol{\xi}_A+\mathbf{b}/2|}{|\mathbf{z}+\mathbf{r}_A-\boldsymbol{\xi}_A+\mathbf{b}/2|^2}\right)\\
&\times\cos\big(k|\mathbf{z}+\mathbf{r}_A-\boldsymbol{\xi}_A-\mathbf{b}/2|-k|\mathbf{z}+\mathbf{r}_A-\boldsymbol{\xi}_A+\mathbf{b}/2|\\
&\qquad+\alpha_M(\boldsymbol{\xi}_A+\mathbf{b}/2,\boldsymbol{\xi}_A-\mathbf{b}/2)\big)\delta(\boldsymbol{\xi}_A).
\end{aligned}
\tag{9}
$$

Each term of $\mathbf{W}_{MD}(\boldsymbol{\xi}_A;\mathbf{r}_A)$ consists in the overlapping of two cones. The geometric factor of the first term corresponds to a cone with vertex at the respective pinhole and Lorentzian shaped base over the detection area at the D-plane. The geometric factor of the second term is the same for $\mathbf{W}_{MD}^{(+)}(\boldsymbol{\xi}_A;\mathbf{r}_A)$ and $\mathbf{W}_{MD}^{(-)}(\boldsymbol{\xi}_A;\mathbf{r}_A)$, and describes a cone with vertex at the midpoint between the pinholes, and cosine shaped base with Lorentzian envelope over the detection area at the D-plane. The weights of these cones, as they overlap, depend on the values of $S_M(\boldsymbol{\xi}_A)$ and $|\mu_M(\boldsymbol{\xi}_A+\mathbf{b}/2,\boldsymbol{\xi}_A-\mathbf{b}/2)|$ respectively. Therefore, it is possible to maintain $S_M(\mathbf{b}/2)=S_M(-\mathbf{b}/2)$ and to change $|\mu_M(\mathbf{b})|$ by changing the pinhole separation $\mathbf{b}$.

It can be shown that the cones $\mathbf{W}_{MD}^{(\pm)}(\boldsymbol{\xi}_A;\mathbf{r}_A)$ result from the modulation of the cone of the first term due to the cone of the second term. Because the cone due to the first term of $\mathbf{W}_{MD}^{(\pm)}(\boldsymbol{\xi}_A;\mathbf{r}_A)$ is associated with the energy that crosses each pinhole, it is referred to as *real cone* while the cone described by the second term is named, *modulating cone* (Fig. 2). Examples of the geometric factors of the cones $\mathbf{W}_{MD}^{(+)}(\boldsymbol{\xi}_A;\mathbf{r}_A)$, $\mathbf{W}_{MD}^{(-)}(\boldsymbol{\xi}_A;\mathbf{r}_A)$, and $\mathbf{W}_{MD}(\boldsymbol{\xi}_A;\mathbf{r}_A)$ for light and for single massive particles, are shown, near the M-plane, in Fig. 4, and at a larger distance, in Fig. 5. The corresponding cones for light and particles have closely similar geometries but differ in scale. Eqs. (3a) and (9) yield $S_D(\mathbf{r}_A) = S_D^{(+)}(\mathbf{r}_A) + S_D^{(-)}(\mathbf{r}_A)$, with

$$\begin{aligned} S_D^{(\pm)}(\mathbf{r}_A) = & \left(\frac{k}{4\pi}\right)^2 S_M(\pm\mathbf{b}/2)\left(\frac{z+|\mathbf{z}+\mathbf{r}_A \pm \mathbf{b}/2|}{|\mathbf{z}+\mathbf{r}_A \pm \mathbf{b}/2|^2}\right)^2 \\ & + \left(\frac{k}{4\pi}\right)^2 |\mu_M(\mathbf{b})|\sqrt{S_M(\mathbf{b}/2)}\sqrt{S_M(-\mathbf{b}/2)} \\ & \times \left(\frac{z+|\mathbf{z}+\mathbf{r}_A - \mathbf{b}/2|}{|\mathbf{z}+\mathbf{r}_A - \mathbf{b}/2|^2}\right)\left(\frac{z+|\mathbf{z}+\mathbf{r}_A + \mathbf{b}/2|}{|\mathbf{z}+\mathbf{r}_A + \mathbf{b}/2|^2}\right) \\ & \times \cos\big(k|\mathbf{z}+\mathbf{r}_A - \mathbf{b}/2| - k|\mathbf{z}+\mathbf{r}_A + \mathbf{b}/2| + \alpha_M(\mathbf{b})\big) \end{aligned} \tag{10}$$

the energy distributions along the cross-sections of the $\mathbf{W}_{MD}^{(\pm)}(\boldsymbol{\xi}_A; \mathbf{r}_A)$ cones in graphs (A) to (D) of Figs. 4 and 5. Vertical profiles are examples at given distances z. The condition $S_D^{(\pm)}(\mathbf{r}_A) \geq 0$ means that the magnitude of the negative value of the cosine term in Eq. (10), at a given point, is not greater than the value of the first term at the same point. This relation is fulfilled at all points in the regions delimited by the dotted lines near the M-plane in graphs (A) and (D) of Fig. 4, while it is extended all over the MD-stage volume in far-field condition (graphs (A) to (D) in Fig. 5). By removing the cosine term, (i.e. the correlation term), only the real cones, given by the first term of Eq. (10) occupy the whole stage volume. Thus, the regions for which $S_D^{(\pm)}(\mathbf{r}_A) < 0$ holds are forbidden for the propagation of waves or particles in the MD-stage, while $S_D^{(\pm)}(\mathbf{r}_A) \geq 0$ gives the physical condition for propagation.

The cones in graphs (E) and (F) in Figs. 4 and 5 result from the overlap of the cones in graphs (A and C) and (B and D) respectively. Their cross-sections determine the energy distribution $S_D(\mathbf{r}_A)$, exemplified by the vertical profiles in the graphs for a given distance z. It fulfills the condition $S_D(\mathbf{r}_A) \geq 0$ in the whole volume of the MD-stage. Therefore, interference of waves and of single particles is described in the same way and accurately predicted by Eq. (10) under the condition $S_D^{(\pm)}(\mathbf{r}_A) \geq 0$.

If $|\mathbf{b}| \to \sigma$ by maintaining $|\mathbf{b}| < \varepsilon$ then $|\mu_M(\mathbf{b})|$ diminishes and, as a consequence, also the fringe contrast in Figs. 4 and 5 decreases. This condition can also be realized, even without changing the fringe geometry, by reducing the length z' of the SM-stage. This result reveals the different physical meanings of the terms of $S_D^{(\pm)}(\mathbf{r}_A)$. The first term in Eq. (10) is a real and positive-definite Lorentzian-like function, whose values are inde-

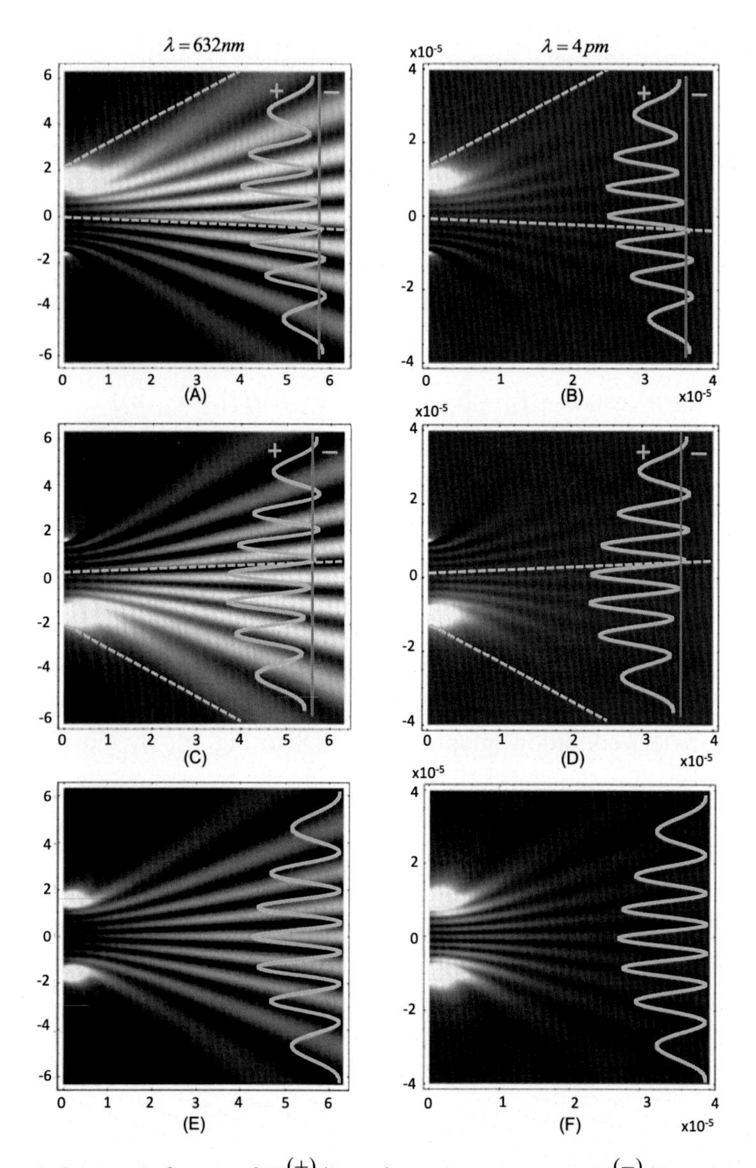

Figure 4 Geometric factors of $\mathbf{W}_{MD}^{(+)}(\boldsymbol{\xi}_A;\mathbf{r}_A)$ on the upper row, $\mathbf{W}_{MD}^{(-)}(\boldsymbol{\xi}_A;\mathbf{r}_A)$ on the mid-row, and $\mathbf{W}_{MD}(\boldsymbol{\xi}_A;\mathbf{r}_A)$ on the bottom row, near the M-plane, for light waves (left column) and single massive particles (right column) and under the conditions $S_M(\pm b/2) = S_0$, $\mu_M(b) = 1$, and $b = 5\lambda$. The vertices of the corresponding real cones are respectively at $\xi_A = b/2, -b/2, \pm b/2$. In all graphs the horizontal axis is z and units of all the axes are in μm. The vertical axes are ξ_A at $z = 0$ and r_A otherwise. The vertical graphs show the energy profiles through the cross section at the right end of each frame. Dotted lines in (A) to (D) delimit the region in which the real cone configuration is changed by the modulating cone. Graphs were enhanced for presentation purposes.

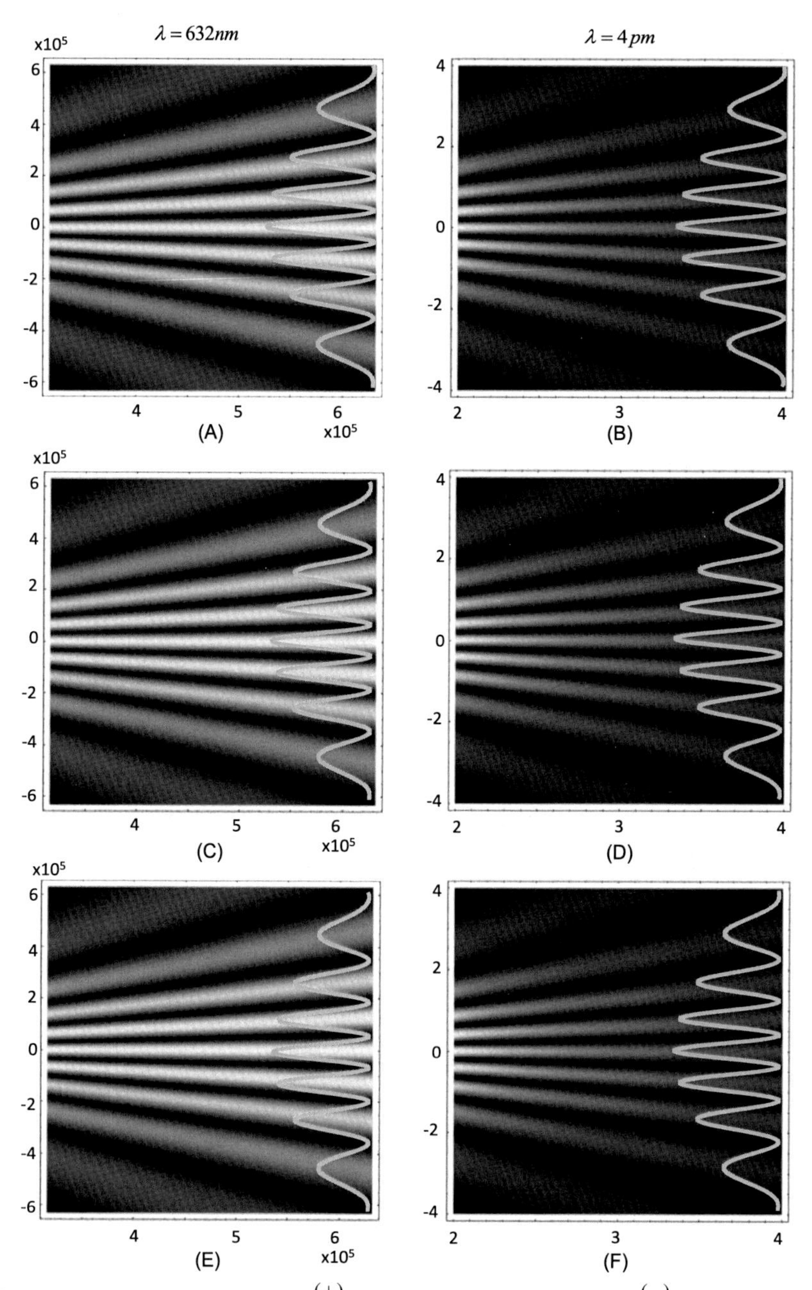

Figure 5 Geometric factors of $\mathbf{W}_{MD}^{(+)}(\boldsymbol{\xi}_A; \mathbf{r}_A)$ on the upper row, $\mathbf{W}_{MD}^{(-)}(\boldsymbol{\xi}_A; \mathbf{r}_A)$ on the mid-row, and $\mathbf{W}_{MD}(\boldsymbol{\xi}_A; \mathbf{r}_A)$ on the bottom row, far from the M-plane, corresponding to those in Fig. 4. Graphs were enhanced for presentation purposes. In all graphs the horizontal axis is z and units of all the axes are in μm.

pendent from the two-point correlation at the M-plane. It is provided by the energy of the wave or of the particle that crosses any of the two pinholes, $S_M(\boldsymbol{\xi}_A = \pm\mathbf{b}/2)$, and therefore it is a physical observable. The second term in Eq. (10) closely depends on the two-point correlation at the M-plane, its geometry arises by the MD-stage configuration and takes positive and negative real values, so that it is not strictly a physical observable. We name it the *geometric potential* (Castañeda, 2017b) because its modulating behavior drives waves and particles towards the bright fringes in graphs of Figs. 4 and 5.

It should be emphasized that the geometric potential is not associated to an external physical agent as, for instance, the scalar potential term included in the Schrödinger equation to account for particles moving through an electric field. For waves and particles moving in a field free region, therefore, the time-independent Schrödinger equation reduces to the Helmholtz equation $\nabla^2\psi(\mathbf{r}) + \frac{2mE}{\hbar^2}\psi(\mathbf{r}) = 0$ which clearly does not contain a potential term. In order to deduce a geometric potential, two Helmholtz equations, referred to positions $\mathbf{r}_\pm$ in center and difference coordinates system and coupled by the eigen-value $-2mE/\hbar^2$, are needed. The solution of this equation system by the Green's function method shows that a geometric potential is present in the non-paraxial modal expansion for the two-point correlation described by Eqs. (1).

Moreover, it must be noted that all the graphs show that the spatial frequency of the fringes diminishes and the fringe intensity decays monotonically with increasing distance from the optical axis. The decay of the cones along the z-axis follows the $1/z^2$-law (Born & Wolf, 1993). It is shown in the patterns for single massive particles but removed for presentation purposes in the patterns for light.

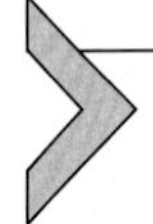

4. PECULIARITIES OF THE NEW GENERAL LAW OF INTERFERENCE

In this section, the main differences are discussed between the general law of interference, synthesized by Eq. (10), and the conventional law (Born & Wolf, 1993). Let us consider the cones $\mathbf{W}_{MD}^{(\pm)}(\boldsymbol{\xi}_A; \mathbf{r}_A)$. Each cone results from the overlap of the real cone with vertex at the corresponding mask pinhole and the modulating cone with vertex at the midpoint between the pinholes. The cross-section of the real cone at the D-plane gives the first term of the corresponding $S_D^{(\pm)}(\mathbf{r}_A)$, while the cross-section of the modulating cone shows the effect of the geometric potential at that plane. This analysis leads us to postulate the existence of point sources of different nature placed, respectively, at the vertices of the real and of the modulating cones so as to produce the respective energy terms of $S_D^{(\pm)}(\mathbf{r}_A)$. A *real point emitter* (labeled $\mathbf{r}$ in Fig. 2) represents the energy source of the wave disturbance or of a point of passage for particles going through the pinholes. A *virtual point emitter* (labeled $\mathbf{v}$ in Fig. 2) represents the source of the geometric potential energy provided by the setup. This energy affects the propagation of waves or particles in the volume of the MD-stage and determines the interference patterns at the detector. Therefore, Eq. (10) describes the (non-local) interaction between each real point emitter and the virtual point emitter at the M-plane. Such interaction constitutes the *new interference principle*. Some features are remarkable, i.e.

(i) The interactions described by $S_D^{(+)}(\mathbf{r}_A)$ and $S_D^{(-)}(\mathbf{r}_A)$ occur separately even when there are wave disturbances or particles crossing both pinholes at the same time. So, the buildup of the interference pattern results from the addition of all occurred interaction events.

(ii) The interactions require that both pinholes remain open when waves or particles (even series of single particles) go through them. When one of the holes is closed, $\mu_M(\mathbf{b}) = 0$ so that the second term of Eq. (10) vanishes. The virtual point source is washed out and, as a consequence, also the origin of the geometric potential.

(iii) The interaction between real and virtual point emitters is conservative. Indeed, the geometric potential vanishes for $\boldsymbol{\xi}_D = 0$ by definition. Furthermore, it has been proved that $\int_D d^2r_A\, \Phi_{MD}(\boldsymbol{\xi}_A; \mathbf{r}_A; k, z) = 1$ and $\int_D d^2r_A\, \Phi_{MD}(\boldsymbol{\xi}_+, \boldsymbol{\xi}_-; \mathbf{r}_A; k, z) = 0$ (Castañeda, 2014; Castañeda et al., 2016a), so that Eq. (3a) yields $\int_D d^2r_A\, S_D(\mathbf{r}_A) = \int_M d^2\xi_A \int_D d^2r_A\, \mathbf{W}_{MD}(\boldsymbol{\xi}_A; \mathbf{r}_A) = \int_M d^2\xi_A\, S_M(\boldsymbol{\xi}_A)$.

Other differences between the conventional description of interference and the present one are worth mentioning. According to the standard description, the acceptance of the wave superposition implies that interference fringes are produced only if the wave-front or the quantum wave function are spread over the mask plane of Fig. 2 and portions of them are considered to describe the crossing events at both pinholes. On the contrary, with our model, interference is explained without taking into account the interaction between two portions of an original wave front emerging from the two pinholes, or mysterious assumptions such as 'delocalized particles', 'self-interference', and 'wave collapse' in case of material particles going through the pinholes one at a time. As a consequence, also the wave superposition principle is not needed.

Without resorting to the standard theoretical description, although powerful, the new law of interference explains carefully the existing experimental results. Let us consider a double-slit, electron interference experiment in which the hits of single particles at the detector are recorded in individual frames. Subsequently all the frames are overlapped and an interference pattern is obtained (see Frabboni et al., 2012; Matteucci, 2013; Matteucci et al., 2013 and references therein). In general, the process to obtain a fringe system can be directly described as the action of the geometric potential on individual particles independently of the pinhole they have crossed. The present interpretation applied to electron interference is illustrated with the simulation movies available at the following YouTube sites:

https://youtu.be/gcKUWLjXvBQ
https://youtu.be/R4zBLL1Wv10
https://youtu.be/wgCb7O9eUqE

We point out, once more, that the observed results are explained without resorting to the wave–particle dualism and to the superposition principle. In addition, the new general law of interference leads to a peculiar prediction which, as far as we know, has never been reported.

Let us consider that the two pinholes of Fig. 2 are open and that the requirements of the two-point correlation are fulfilled. In these conditions, a given number of events crossing one of the pinholes builds up a fringe pattern at the D-plane independently of the events taking place at the other pinhole. In fact, the simulations in far-field condition reported in Fig. 5, show that an interference pattern results from the addition of two subsets of crossing events, pertaining to the two pinholes, which produce, individually, the same fringe system.

An experimental proof of this prediction is very demanding. It implies the possibility, to record the buildup of all the events coming from one of the holes by keeping, at the same time, both pinholes open. A possible demonstration of this intriguing aspect of the new law of interference might be to use the weak measurement technique (Kocsis et al., 2011; Mahler et al., 2016).

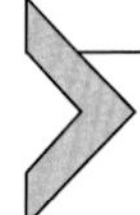

5. EXTENDED SOURCES AND INTERFERENCE FROM GRATINGS

The new general law of interference is applicable to any experiment, including those prepared with extended effective sources and realized with a grating. In this context, some features should be taken into account, i.e. (i) in the SM-stage, correlation cones produced by extended effective sources have a non-Lorentzian envelope and a narrower aperture angle in comparison to that produced by a point source. As a consequence, the structured supports at the M-plane are smaller and therefore, pairs of points on this plane are, in general, partially correlated. (ii) Several correlation cones are required in order to cover all the possible pairs of points selected by the grating slits. (iii) A virtual point emitter is located at the midpoint of the structured support of each cone at the M plane. (iv) Once the extended source is put into operation, a real point emitter is considered at each grating slit, so that a set of real point emitters distributed on the M-plane represents the slit. Therefore, any specific experiment is characterized by discrete and finite sets of real and virtual point emitters at the M-plane and the corresponding sets of non-paraxial modes for the kernel.

In order to describe interference experiments in the framework of the new interpretation, let us introduce the mathematical tool named the *spectrum of classes of point emitters* (Castañeda & Muñoz, 2016). It determines the interaction map, between real and virtual point emitters, that characterizes the specific interference experiment (Castañeda, 2017b). The spectrum of classes of point emitters is defined by the coefficient of the modal expansion, in Eq. (3b), which takes the values $S_M(\boldsymbol{\xi}_A)$ denoting the real point emitter at a given position $\boldsymbol{\xi}_A$, and by $|\mu_M(\boldsymbol{\xi}_+, \boldsymbol{\xi}_-)|\sqrt{S_M(\boldsymbol{\xi}_+)}\sqrt{S_M(\boldsymbol{\xi}_-)}$ which gives the component of the class with separation vector $\boldsymbol{\xi}_D$ of the virtual point emitter at certain position $\boldsymbol{\xi}_A$, that can be shared with a real point emitter (Castañeda & Muñoz, 2016). So, we obtain

$$\mathbf{SC}_M(\boldsymbol{\xi}_A, \boldsymbol{\xi}_D) = \left\{S_M(\boldsymbol{\xi}_A), |\mu_M(\boldsymbol{\xi}_+, \boldsymbol{\xi}_-)|\sqrt{S_M(\boldsymbol{\xi}_+)}\sqrt{S_M(\boldsymbol{\xi}_+)}\right\}, \tag{11a}$$

with

$$\mathbf{SC}_M^{(R)}(\boldsymbol{\xi}_A, 0) = \left\{S_M(\boldsymbol{\xi}_A)\right\} \tag{11b}$$

the zeroth order class composed only by the subset of real point emitters at the grating slits, as the superscript R indicates, and

$$\mathbf{SC}_M^{(V)}(\boldsymbol{\xi}_A, \boldsymbol{\xi}_D^{(n)}) = \left\{|\mu_M(\boldsymbol{\xi}_+^{(n)}, \boldsymbol{\xi}_-^{(n)})|\sqrt{S_M(\boldsymbol{\xi}_+^{(n)})}\sqrt{S_M(\boldsymbol{\xi}_-^{(n)})}\right\} \tag{11c}$$

the n-order class composed by the pairs of pinholes of separation $\boldsymbol{\xi}_D^{(n)}$, $n = 1, 2, \ldots$, ordered by the condition $|\boldsymbol{\xi}_D^{(n)}| < |\boldsymbol{\xi}_D^{(n+1)}|$. The members of the n-order class distribute over the grating length at the rate of one member with midpoint at a given position $\boldsymbol{\xi}_A$. It means that (i) there is only one member of a given class in any structured support, and (ii) the members of different classes inscribed in the structured support centered at a given $\boldsymbol{\xi}_A$ contribute only to the virtual point emitter at the support center, as indicated by the superscript V (Castañeda & Muñoz, 2016).

The discreteness of the zeroth order class is crucial to characterize extended effective sources and gratings (Castañeda & Muñoz, 2016). Specifically, the minimum spacing of the zeroth order class that assures this characterization must be defined. First of all, let us consider wave interference. The following conditions have been established for a zeroth order class composed by N real point emitters with spacing b under full two-point correlation (Castañeda & Muñoz, 2016; Castañeda, 2017a), i.e.

$$\begin{array}{ll}
b > \lambda & \text{interference,} \\
\lambda/10 < b \leq \lambda < (N-1)b & \text{diffraction,} \\
(N-1)b \leq \lambda & \text{transition,} \\
(N-1)b \leq \lambda/10 & \text{single real point emitter.}
\end{array}$$

The condition above referred to as 'transition', means that bell-like patterns different from the Lorentzian ones are produced at the D-plane. The condition for single real point emitter is independent from the value of N and the two-point correlation at the M-plane. So, the characterization of extended effective sources of waves should fulfill the second condition while the set that represents the grating stands the first condition.

Fig. 6 shows the effect of the effective source size on the experiment preparation. Thermal, uniform effective wave sources, of size $(N-1)\lambda/2$, provide the profiles of the correlation degrees and energy distributions at the M-plane. The term uniform means that the real point emitters are assumed identical. Profiles in (A) show the expected reduction in size of the structured support (σ) as the size of the effective source of waves increases. Nevertheless, for a given source-grating distance z', such extended, uncorrelated effective sources produce Lorentzian-like distributions of energy at the M-plane. To understand the fundamentals of the construction of the interaction map using the spectrum of classes of point emitters, let us consider a grating with three collinear, equidistant pinholes with period $|\mathbf{b}|$, at the M-plane, so that

$$\begin{aligned}
t_M(\boldsymbol{\xi}_+)t_M(\boldsymbol{\xi}_-) = {} & \left[\delta(\boldsymbol{\xi}_A + \mathbf{b}) + \delta(\boldsymbol{\xi}_A) + \delta(\boldsymbol{\xi}_A - \mathbf{b})\right]\delta(\boldsymbol{\xi}_D) \\
& + \left[\delta(\boldsymbol{\xi}_D + \mathbf{b}) + \delta(\boldsymbol{\xi}_D - \mathbf{b})\right]\left[\delta(\boldsymbol{\xi}_A + \mathbf{b}/2) + \delta(\boldsymbol{\xi}_A - \mathbf{b}/2)\right] \\
& + \left[\delta(\boldsymbol{\xi}_D + 2\mathbf{b}) + \delta(\boldsymbol{\xi}_D - 2\mathbf{b})\right]\delta(\boldsymbol{\xi}_A).
\end{aligned} \tag{12}$$

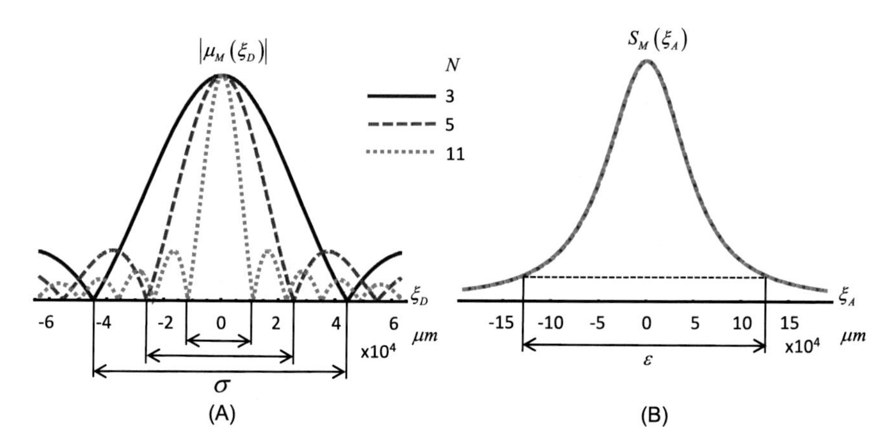

Figure 6 (A) Profiles of the correlation degrees at the M-plane ($z' = 10^5\lambda$, $\lambda = 632$ nm) generated by an effective source of N uncorrelated and identical real point emitters with period $\lambda/2$. (B) All the effective sources, disregarding their sizes, produce the same Lorentzian distribution of energy at the M-plane. Although the size of the structured support σ decreases as the number of point emitters increases, the size of the illuminated area ε remains the same.

Eq. (12) determines the maximum number of classes of point emitters and their distribution at the M-plane. Three classes of point emitters are defined, (i) the zeroth order class with three real point emitters at the pinhole positions $\boldsymbol{\xi}_A = 0, \pm\mathbf{b}$, (ii) the first order class $\boldsymbol{\xi}_D = \pm\mathbf{b}$ with two members that determine the virtual point emitters at the positions $\boldsymbol{\xi}_A = \pm\mathbf{b}/2$ and, (iii) the second order class $\boldsymbol{\xi}_D = \pm 2\mathbf{b}$, with one member that determines the virtual point emitter at the position $\boldsymbol{\xi}_A = 0$. This arrangement is sketched in Fig. 7 in case of full two-point correlation, with $|\mathbf{b}| = 3\lambda$ and $\lambda = 632$ nm.

According to Eqs. (11) the corresponding spectrum of classes is given by

$$\mathbf{SC}_M^{(R)}(\boldsymbol{\xi}_A, 0) = \{S_M(\mathbf{b}), S_M(0), S_M(-\mathbf{b})\} \tag{13a}$$

for the zeroth order class, whose terms are represented by the three white spots in Fig. 7(A),

$$\mathbf{SC}_M^{(V)}(\pm\mathbf{b}/2, \mathbf{b}) = \{|\mu_M(\pm\mathbf{b}, 0)|\sqrt{S_M(\pm\mathbf{b})}\sqrt{S_M(0)}\}, \tag{13b}$$

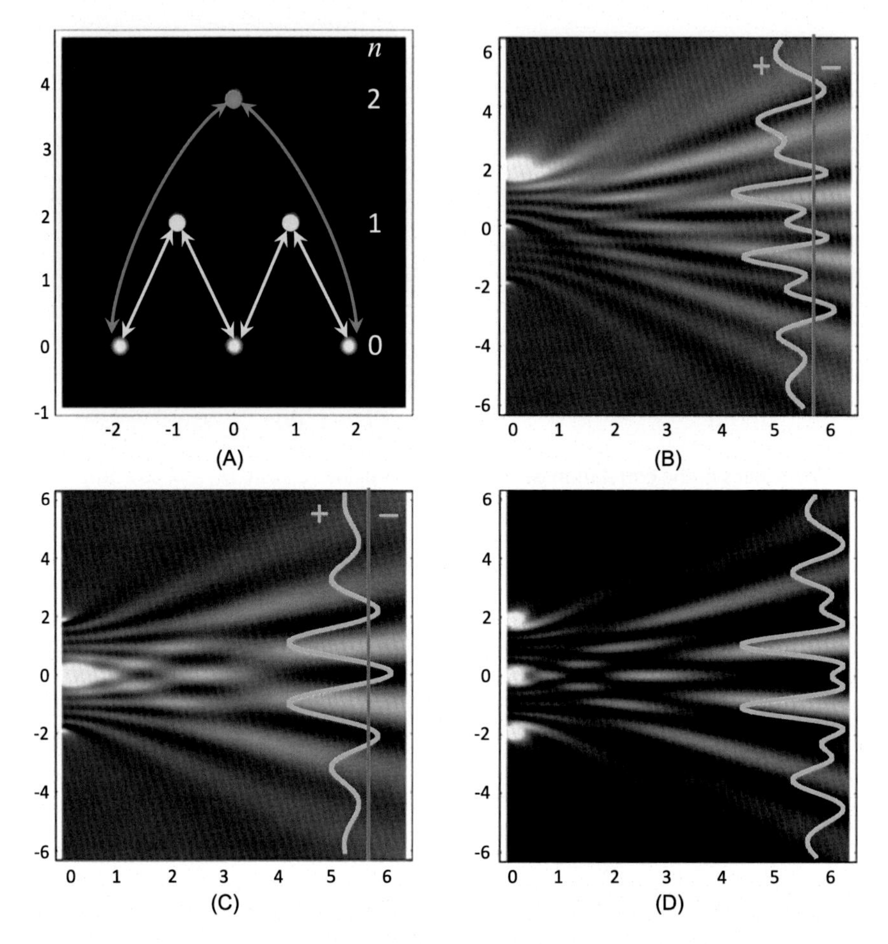

Figure 7 Application of the interaction principle to wave interference. The spectrum of classes of point emitters determined by Eqs. (13) is presented in (A), with n the class order. The horizontal ξ_A and vertical ξ_D axes are in μm. The arrows link the interacting point emitters. The $\mathbf{W}_{MD}^{(+)}(\xi_A;\mathbf{r}_A)$ and $\mathbf{W}_{MD}^{(0)}(\xi_A;\mathbf{r}_A)$ cones near the M-plane are shown in (B) and (C) respectively. Vertical profiles describe the interaction energies $S_D^{(+)}(\mathbf{r}_A)$ and $S_D^{(0)}(\mathbf{r}_A)$ at $z=10\lambda$. The overall energy cone $\mathbf{W}_{MD}(\xi_A;\mathbf{r}_A)$ is shown in (D) and its vertical profile describes the interference pattern recorded by a detector at the D-plane at $z=10\lambda$. In (B) to (D) the horizontal axis, z and the vertical axes, ξ_A at $z=0$ and r_A for $z>0$, are in μm. Graphs are enhanced for presentation purposes.

containing the two virtual point emitters of the first order class, shown with yellow spots in Fig. 7(A), and

$$\mathbf{SC}_M^{(V)}(0,2\mathbf{b})=\left\{\left|\mu_M(\mathbf{b},-\mathbf{b})\right|\sqrt{S_M(\mathbf{b})}\sqrt{S_M(-\mathbf{b})}\right\} \tag{13c}$$

with the virtual point emitter of the second order class, blue spot in Fig. 7(A).

Each dot in graph (A) is the member of the class of separation vector $\boldsymbol{\xi}_D$ (vertical coordinate axis) placed at the position $\boldsymbol{\xi}_A$ (horizontal coordinate axis). The central, real point emitter interacts only with the virtual point emitters of the first class, as shown by the yellow arrows, while the other two real point emitters interact with the virtual point emitters of the first and second classes as shown by the blue arrows. The interaction between the central real and virtual point emitters is forbidden. Thus, Eqs. (3b), (3c), and (12) determine cones in the MD-stage that describe the effect of the overall interaction, depicted in Fig. 7(A), for the propagation of waves and single particles. These cones are given by

$$\begin{aligned}
\mathbf{W}_{MD}^{(\pm)}(\boldsymbol{\xi}_A;\mathbf{r}_A) &= 2\left(\frac{k}{4\pi}\right)^2 S_M(\boldsymbol{\xi}_A)\left(\frac{z+|\mathbf{z}+\mathbf{r}_A-\boldsymbol{\xi}_A|}{|\mathbf{z}+\mathbf{r}_A-\boldsymbol{\xi}_A|^2}\right)^2 \delta(\boldsymbol{\xi}_A\mp\mathbf{b}) \\
&+\left(\frac{k}{4\pi}\right)^2 \sqrt{S_M(\boldsymbol{\xi}_A+\mathbf{b}/2)}\sqrt{S_M(\boldsymbol{\xi}_A-\mathbf{b}/2)} \\
&\times \left|\mu_M(\boldsymbol{\xi}_A+\mathbf{b}/2,\boldsymbol{\xi}_A-\mathbf{b}/2)\right|\delta(\boldsymbol{\xi}_A\mp\mathbf{b}/2) \\
&\times\left(\frac{z+|\mathbf{z}+\mathbf{r}_A-\boldsymbol{\xi}_A-\mathbf{b}/2|}{|\mathbf{z}+\mathbf{r}_A-\boldsymbol{\xi}_A-\mathbf{b}/2|^2}\right)\left(\frac{z+|\mathbf{z}+\mathbf{r}_A-\boldsymbol{\xi}_A+\mathbf{b}/2|}{|\mathbf{z}+\mathbf{r}_A-\boldsymbol{\xi}_A+\mathbf{b}/2|^2}\right) \\
&\times\cos\big(k|\mathbf{z}+\mathbf{r}_A-\boldsymbol{\xi}_A-\mathbf{b}/2|-k|\mathbf{z}+\mathbf{r}_A-\boldsymbol{\xi}_A+\mathbf{b}/2| \\
&\qquad+\alpha_M(\boldsymbol{\xi}_A+\mathbf{b}/2,\boldsymbol{\xi}_A-\mathbf{b}/2)\big) \\
&+\left(\frac{k}{4\pi}\right)^2 \sqrt{S_M(\boldsymbol{\xi}_A+\mathbf{b})}\sqrt{S_M(\boldsymbol{\xi}_A-\mathbf{b})} \\
&\times \left|\mu_M(\boldsymbol{\xi}_A+\mathbf{b},\boldsymbol{\xi}_A-\mathbf{b})\right|\delta(\boldsymbol{\xi}_A) \\
&\times\left(\frac{z+|\mathbf{z}+\mathbf{r}_A-\boldsymbol{\xi}_A-\mathbf{b}|}{|\mathbf{z}+\mathbf{r}_A-\boldsymbol{\xi}_A-\mathbf{b}|^2}\right)\left(\frac{z+|\mathbf{z}+\mathbf{r}_A-\boldsymbol{\xi}_A+\mathbf{b}|}{|\mathbf{z}+\mathbf{r}_A-\boldsymbol{\xi}_A+\mathbf{b}|^2}\right) \\
&\times\cos\big(k|\mathbf{z}+\mathbf{r}_A-\boldsymbol{\xi}_A-\mathbf{b}|-k|\mathbf{z}+\mathbf{r}_A-\boldsymbol{\xi}_A+\mathbf{b}| \\
&\qquad+\alpha_M(\boldsymbol{\xi}_A+\mathbf{b},\boldsymbol{\xi}_A-\mathbf{b})\big) \qquad (14a)
\end{aligned}$$

that delineates the interactions between the real point emitters at $\boldsymbol{\xi}_A = \pm\mathbf{b}$ and the virtual point emitters of the first and the second order classes. Fig. 7(B) shows the cone $\mathbf{W}_{MD}^{(+)}(\boldsymbol{\xi}_A;\mathbf{r}_A)$ near the M-plane while the ($\mathbf{W}_{MD}^{(-)}(\boldsymbol{\xi}_A;\mathbf{r}_A)$ cone is omitted in Fig. 7 because it is symmetric of $\mathbf{W}_{MD}^{(+)}(\boldsymbol{\xi}_A;\mathbf{r}_A)$ with respect to the axis $\boldsymbol{\xi}_A = \mathbf{r}_A = 0$. In addition,

$$
\begin{aligned}
\mathbf{W}^{(0)}_{MD}(\boldsymbol{\xi}_A;\mathbf{r}_A) &= 2\left(\frac{k}{4\pi}\right)^2 S_M(\boldsymbol{\xi}_A)\left(\frac{z+|\mathbf{z}+\mathbf{r}_A-\boldsymbol{\xi}_A|}{|\mathbf{z}+\mathbf{r}_A-\boldsymbol{\xi}_A|^2}\right)^2 \delta(\boldsymbol{\xi}_A)\\
&+\left(\frac{k}{4\pi}\right)^2 \sqrt{S_M(\boldsymbol{\xi}_A+\mathbf{b}/2)}\sqrt{S_M(\boldsymbol{\xi}_A-\mathbf{b}/2)}\\
&\times\left|\mu_M(\boldsymbol{\xi}_A+\mathbf{b}/2,\boldsymbol{\xi}_A-\mathbf{b}/2)\right|\left[\delta(\boldsymbol{\xi}_A+\mathbf{b}/2)+\delta(\boldsymbol{\xi}_A-\mathbf{b}/2)\right]\\
&\times\left(\frac{z+|\mathbf{z}+\mathbf{r}_A-\boldsymbol{\xi}_A-\mathbf{b}/2|}{|\mathbf{z}+\mathbf{r}_A-\boldsymbol{\xi}_A-\mathbf{b}/2|^2}\right)\left(\frac{z+|\mathbf{z}+\mathbf{r}_A-\boldsymbol{\xi}_A+\mathbf{b}/2|}{|\mathbf{z}+\mathbf{r}_A-\boldsymbol{\xi}_A+\mathbf{b}/2|^2}\right)\\
&\times\cos\big(k|\mathbf{z}+\mathbf{r}_A-\boldsymbol{\xi}_A-\mathbf{b}/2|-k|\mathbf{z}+\mathbf{r}_A-\boldsymbol{\xi}_A+\mathbf{b}/2|\\
&\qquad+\alpha_M(\boldsymbol{\xi}_A+\mathbf{b}/2,\boldsymbol{\xi}_A-\mathbf{b}/2)\big)
\end{aligned}
\tag{14b}
$$

describes the interactions between the real point emitter at $\boldsymbol{\xi}_A=0$ and the virtual point emitters of the first order class, Fig. 7(C). Eqs. (3a) and (14a), (14b) yield $S_{MD}(\mathbf{r}_A)=S^{(+)}_{MD}(\mathbf{r}_A)+S^{(0)}_{MD}(\mathbf{r}_A)+S^{(-)}_{MD}(\mathbf{r}_A)$ for the energy distribution of the interference pattern recorded at the D-plane. It is exemplified by the vertical profiles of the cross sections of the $\mathbf{W}_{MD}(\boldsymbol{\xi}_A;\mathbf{r}_A)$ cone in Fig. 7(D). Its components are the interaction energies associated to each real point emitter. Each one has in turn two terms obtained from the application of the general law of interference to the corresponding interactions. It is worth noting that the factor 2, in the first terms of Eqs. (14), points out that a real point emitter has separate interactions with each virtual point emitter.

Interaction energies are exemplified by the vertical profiles of the cross-sections at $z=10\lambda$ of the cones in Fig. 7(B) and (C). The wave energy is driven towards the regions in which the interaction energy takes on positive values (i.e. the bright fringes in cone graphs of Fig. 7). Negative values of the interaction energy denote forbidden regions (i.e. the dark fringes in the graphs of Fig. 7). The resulting cumulative effect over the complete set of interactions determines, for waves or particles, what is called destructive interference.

The effect of the class filtering performed by the correlation degree at the M-plane (Castañeda, 2017b) is illustrated in Fig. 8, by considering $\mu_M(\boldsymbol{\xi}_D)=\exp(-|\boldsymbol{\xi}_D|^2/2\sigma^2)$, with $\sigma=1.5\lambda$ in the example of Fig. 7. The spectrum of classes of point emitters shows that the second order class is dropped out, Fig. 8(A), thus eliminating the contribution to the geometric potential and the effects of the corresponding interactions. Furthermore, the virtual point emitters of the first order class are weaker so that low contrast interference patterns take place, Fig. 8(B), (C), (D). Finally, with $\sigma=0.3\lambda$, point emitters of the first and second order classes are no longer active, Fig. 9(A). So, the geometric potential and the interactions

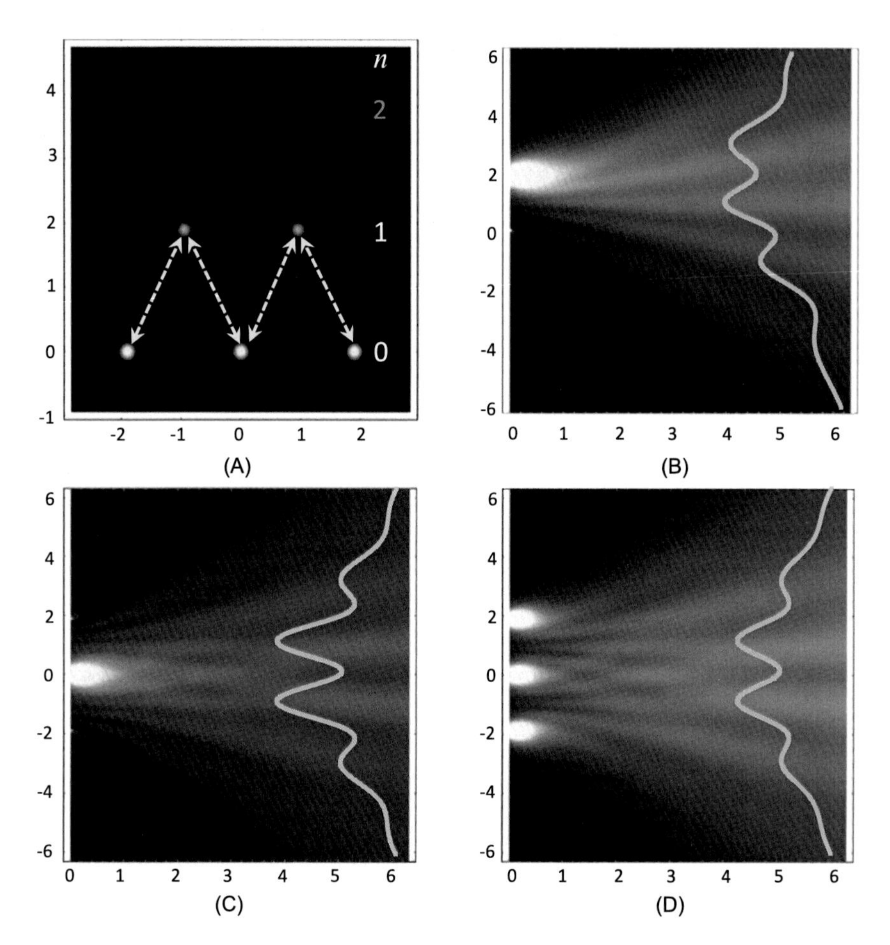

Figure 8 Illustrating the effect of class filtering performed by the correlation degree (Gaussian standard deviation 1.5λ) at the M-plane in the graphs of Fig. 7. (A) Spectrum of classes of point emitters. The second order class is dropped out and the virtual point emitters of the first order class are weaker. Only weak interactions (dotted arrows) between the zero and first order classes take place. As a consequence, low contrast fringes are formed in the cones corresponding to $\mathbf{W}_{MD}^{(+)}(\xi_A; \mathbf{r}_A)$ graph (B), $\mathbf{W}_{MD}^{(0)}(\xi_A; \mathbf{r}_A)$ graph (C), and $\mathbf{W}_{MD}(\xi_A; \mathbf{r}_A)$ graph (D). In (B) to (D) the horizontal axis, z and the vertical axes, ξ_A at $z = 0$ and $\mathbf{r}_A$ for $z > 0$, are in μm. Graphs are enhanced for presentation purposes.

are dropped out. As a consequence, only the Lorentzian real cones associated to the real point emitters determine the configuration of the energy cone in the MD-stage. The recorded pattern does not show interference fringes, as illustrated in Fig. 9(B).

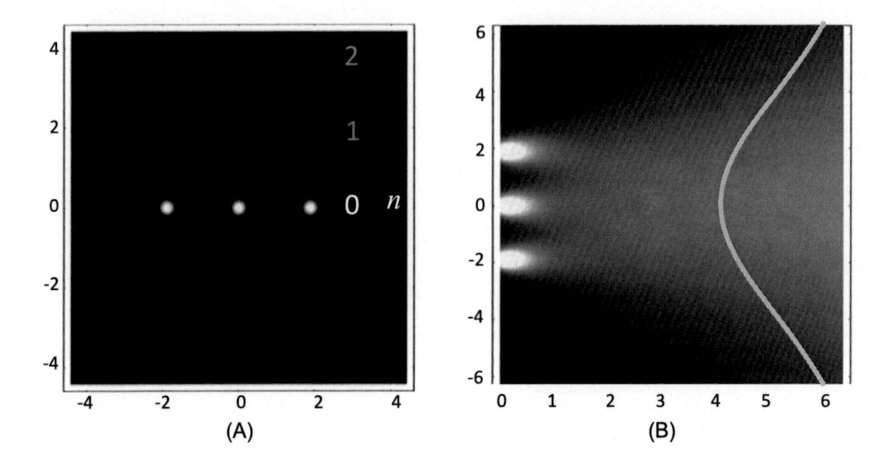

Figure 9 (A) A very low correlation degree at the M-plane (Gaussian standard deviation, 0.3λ) determines, in the MD-stage, an energy cone without interference modulation, (B). In (B) the horizontal axis, z and the vertical axes, $\boldsymbol{\xi}_A$ at $z = 0$ and $\mathbf{r}_A$ for $z > 0$, are in μm. Graphs are enhanced for presentation purposes.

The procedure above constitutes the basic algorithmic element to explain and predict interference with arbitrary number of real point emitters in setups characterized by arbitrary sets of virtual point emitters. In order to exemplify it, let us consider the spectra of classes of point emitters in Fig. 10, column on the left. They characterize the wave interference experiment with a grating consisting of 11 pinholes of period 3λ under Gaussian correlation condition. The maximum set of high order classes, obtained under full two-point correlation, is 10 (i.e. $1 \leq n \leq 10$) with decreasing number of virtual point emitters, i.e. 10, 9, 8, ..., 1 for $n = 1, 2, 3, \ldots, 10$ respectively. The arrows allow the construction of the interaction map, between real and virtual point emitters, in accordance with the following considerations:

(i) The arrow directions are invariant and determine the interaction links. So, they are called the interaction links directions (ILD).

(ii) The bottom head of both arrows should be placed on the selected real point emitter of the zeroth order class, as suggested in Fig. 7(A).

(iii) The selected real point emitter interacts only with the components of the virtual point emitters represented by the dots of the high order classes along the ILD of both arrows. Each interaction is described by the new general law of interference. Here, it is important to consider that all the pairs of points inscribed in the structured

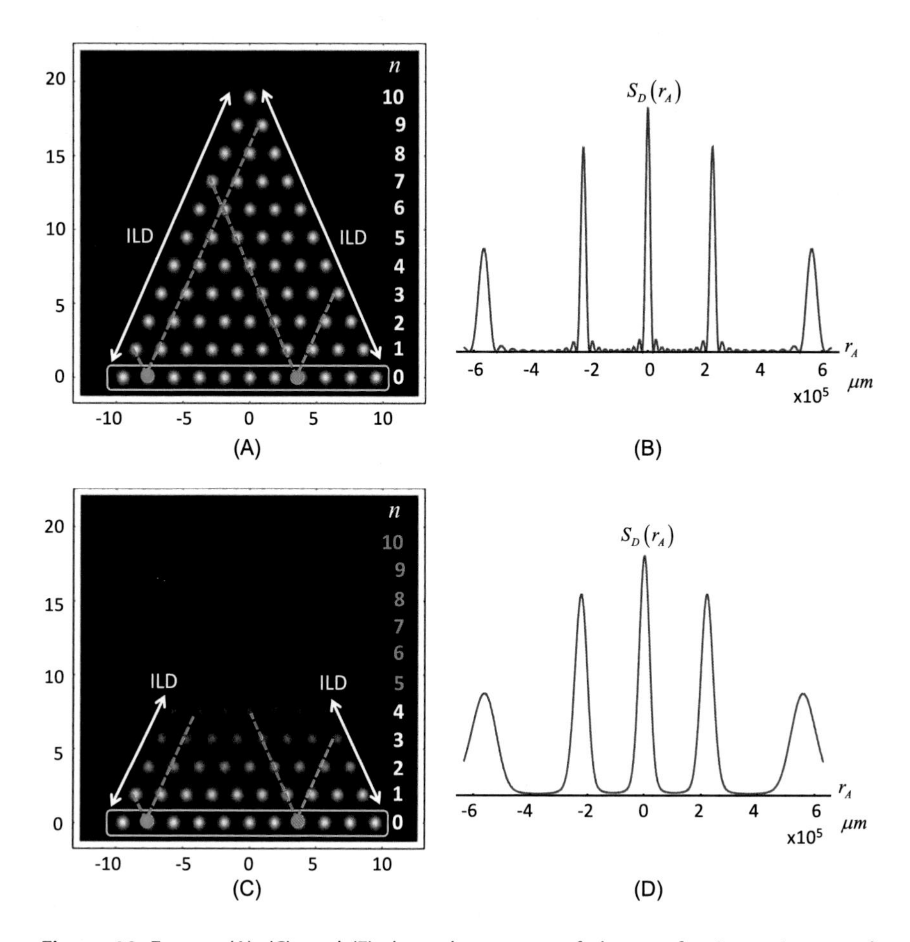

Figure 10 Frames (A), (C), and (E) show the spectra of classes of point emitters and (B), (D), (F) the corresponding far field, ($z = 10^6\lambda$), interference patterns of light ($\lambda =$ 632 nm) obtained with a grating of 11 identical pinholes under Gaussian correlation degree. The zeroth order class is enclosed by the rectangular solid line frame and its real point emitters interact with the components of the virtual point emitters along the ILD-arrows, as indicated by the dotted lines. In (A), the structured support size is much larger than the grating size, 30λ, so that the maximum class order for this experiment is 10. In (B), the interactions produce a high contrasted interference pattern with very narrow main maxima. In (C) and (E), the structured support size is gradually reduced, compared to the grating size, thus filtering high order classes of emitters, respectively. In (D) and (F) are shown the loss of contrast and broadening of the main maxima of the interference patterns corresponding to the conditions (C) and (E), respectively. The units of the horizontal ξ_A-axis and vertical ξ_D-axis in the graphs (A), (C), and (E) are in μm.

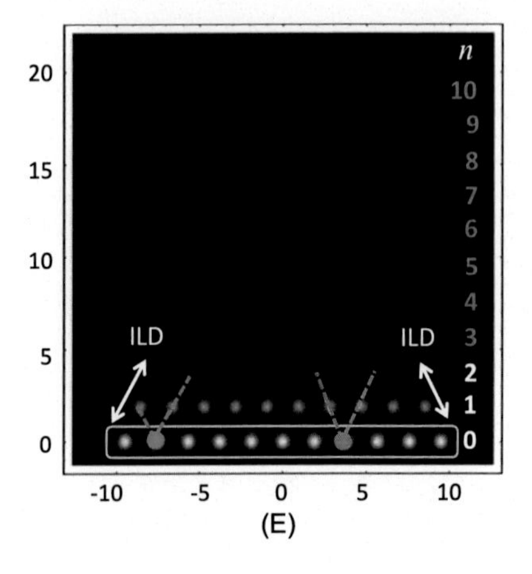

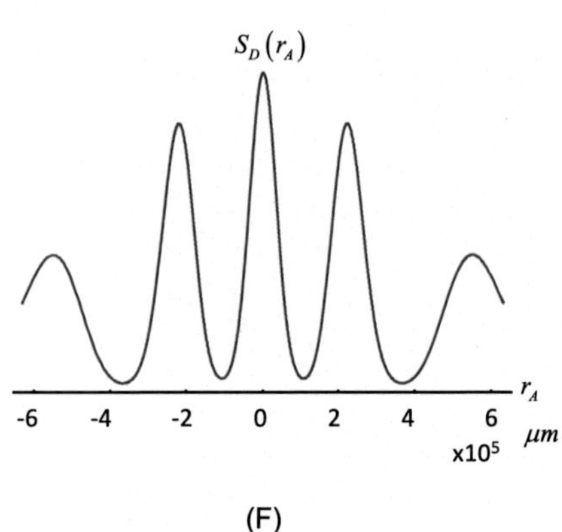

Figure 10 (*continued*)

support centered at a position $\boldsymbol{\xi}_A$ contribute to the virtual point emitter at such position. There is up to one pair with separation vector $\boldsymbol{\xi}_D$ with $|\boldsymbol{\xi}_D| \leq \sigma$ in each structured support (Castañeda & Muñoz, 2016; Castañeda, 2017b), or equivalently up to one member of each class inscribed in the structured support. Such members are therefore the components of the virtual point emitter at the support center.

The complete interaction map is constructed once this procedure is applied to all real point emitters. A Gaussian correlation degree at the M-plane is assumed in the example of Fig. 10. $\sigma \gg 30\lambda$ holds in (A) so that each real point emitter has the maximum number of interactions with virtual point emitters (10 interactions although there are 19 virtual point emitters).

This condition is evidenced by the $S_D(\mathbf{r}_A)$ profile shown in (B). The main maxima are separated by 9, much weaker, secondary maxima. Spatial frequency chirping is shown by the non-uniform angular width and spacing of the main maxima. A Lorentzian envelope modulates the fringe intensity. By increasing step by step the effective source size, the reduced degree of correlation, $\sigma < 30\lambda$ filters some high order classes of emitters as shown in (C) and (E). As a consequence, the number of interactions of each real point emitter reduces. Specifically, classes for $n \geq 4$ in (C) and $n \geq 3$ in (E) are dropped out. Therefore, the corresponding interference profiles (D) and (F) show, (i) an angular broadening of the main maxima, (ii) a washing out of the secondary maxima and, (iii) a decreasing fringe contrast in comparison with the interference pattern obtained with the full two-point correlation condition reported in (B).

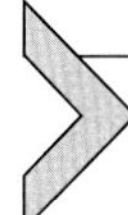

6. DIFFRACTION EFFECTS AND UNCERTAINTY PRINCIPLE

Diffraction experiments of material particles from a single aperture are often used to discuss the close connection between the Heisenberg's uncertainty principle and diffraction effects (Feynman et al., 1965). In case of a monochromatic electron beam impinging on a pinhole, the particle distribution in the far-field is represented by a squared sinc function, calculated with standard optical methods, by assuming a plane probability wave at the entrance surface of the diffracting element (Matteucci, 2011 and references therein). However, the theoretical analysis of the interference fringe modulation of single massive molecules passing through a grating offers a stimulating opportunity for a general study of the relation between particle diffraction and uncertainty principle. Experimental results of interference with particles have been reported, for instance, in Arndt et al. (1999), Bach et al. (2013), Frabboni et al. (2012), Juffmann et al. (2012), Nairz et al. (2003), and Zeilinger et al. (1988).

As a case study we consider the interference of single C_{60} molecules from a grating with nominal constant 100 nm and slit width 55 ± 5 nm, reported by Nairz et al. (2003). In these experiments, a thermal beam of particles, characterized by a limited transversal coherence, has been considered as a plane wave impinging on the grating. With this approach, however, the intensity of the observed interference maxima cannot be accurately calculated from the constructive interference of the wavelets emerging from the grating slits (Castañeda et al., 2016b). As a consequence, also the diffraction curve, which syntheses the complementarity between momentum and position of particles, does not fit with the interference fringes. For this reason, advances of the understanding of the basic connection between diffraction phenomena and the Heisenberg uncertainty principle is of the outmost importance.

Figs. 11(A) and (B) show the profiles of the predicted interference patterns in the framework of our model. Their excellent fit with the experimental data was discussed in detail in Castañeda et al. (2016b) and is not included here. We showed how a Lorentzian profile modulates almost perfectly the interference fringes, a result that cannot be obtained with the standard approach by means of which the molecule distribution at the

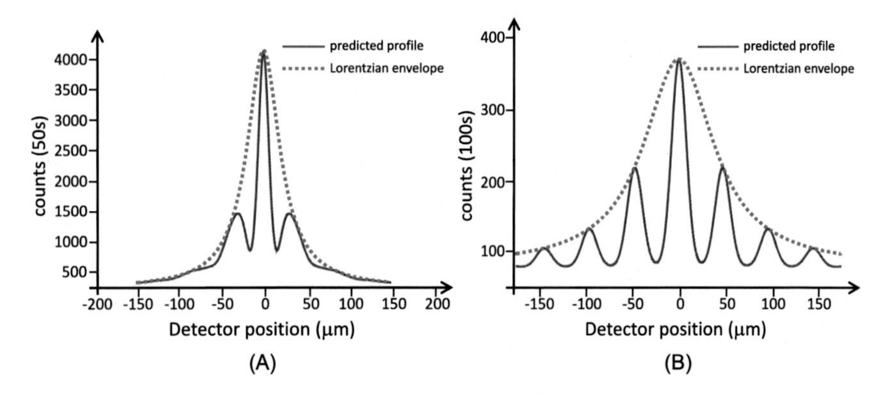

Figure 11 The red solid lines represent the profiles of the interference patterns and the dotted lines show the Lorentzian diffraction envelopes of the interference with single C_{60} molecules, obtained under (A) low two-point correlation and (B) high two-point correlation, reported in Nairz et al. (2003).

detector is calculated by a plane probability wave impinging on the grating.

As discussed in the preceding sections, the Lorentzian diffraction modulation takes place if the grating behaves like an array of single real point emitters with a spacing equal to the grating period, i.e. all possible points of passage of molecules through a slit can be replaced with a single point emitter located at the slit center. In other words, the experimental results indicate that the crossing events of molecules through the points of a slit are not distinguishable and therefore, they produce the same effect of a single real point emitter placed at the slit center. We refer to the width of this region as the uncertainty in position of the passing molecules.

In case of light waves, it has been reported that arrays of real point emitters of length shorter than $\lambda/10$ behave like a single real point emitter placed at the array midpoint, irrespective from the number of emitters in the array and their two-point correlation condition (Castañeda, 2017a). In other words, slits of width narrower than $\lambda/10$ cannot change the angular aperture or the geometry of the diffraction Lorentzian cone. In this sense, $\lambda/10$ represents the uncertainty in position associated to the real point emitters of light waves.

Because of the finite size of the uncertainty in position and its independence of the correlation within the uncertainty region, it is reasonable to model the uncertainty in position as a peculiarity of the geometric potential. For this purpose, let us consider the coefficient of the geometric potential, $\sqrt{S_M(\xi_+)}\sqrt{S_M(\xi_-)}$ in Eq. (10), which is valid for waves or particles going through the interferometric element at the M-plane, Fig. 2. The definition $W_M(\xi_+, \xi_-) = \mu_M(\xi_+, \xi_-)\sqrt{S_M(\xi_+)}\sqrt{S_M(\xi_-)}$ of the two-point correlation at the M-plane assumes implicitly that any pair of points are distinguishable irrespective from their separation ξ_D. Therefore, the geometric potential, provided by the corresponding virtual point emitter, should produce diffraction modulation in the MD-stage.

However, simulations of light wave interference and of observed single molecule interference patterns point out a common behavior of waves and particles, i.e. associated to them, a characteristic length L_C is evaluated that defines the distinguishability of pairs of real point emitters. It means that two real point emitters on the M-plane, for which $|\xi_D| \leq L_C$, are indistinguishable if the geometric potential they contribute does not cause a diffraction modulation effect in the MD-stage. In this condition, the same Lorentzian cone is obtained as that one due to a single real point emitter placed at half distance between the two point emitters. It is equivalent to say that the pair of real point emitters are included in the uncertainty region of the single real point emitter.

By taking into account that $S_M(\xi_A) = W_M(\xi_A, \xi_A)$, it is reasonable to introduce a real valued function $u_M(\xi_D)$ with the following properties, (i) $0 \leq u_M(\xi_D) \leq 1$, (ii) $u_M(0) = 1$, and (iii) $u_M(\xi_D) \doteq 0$ or negligible for $|\xi_D| > L_C$. Furthermore, $u_M(\xi_D)$ should predict the Lorentzian modulation of simulations and fit the experimental observations.

Thus, we introduce a new definition for the two-point correlation at the M-plane which includes the uncertainty in position as a feature of waves or of particles and allows to predict accurately any experimental interference pattern, i.e.

$$W_M(\xi_+, \xi_-) = \begin{cases} \mu_M(\xi_+, \xi_-)\sqrt{S_M(\xi_+)}\sqrt{S_M(\xi_-)} & \text{for} |\xi_D| > L_C \\ S_M(\xi_A)u_M(\xi_D) & \text{for} |\xi_D| \leq L_C \end{cases}. \tag{15}$$

Without lack of generality, let us consider the interference experiment with single C_{60} molecules reported in Nairz et al. (2003). It was designed for far field conditions, which approximates the non-paraxial kernel to a Fourier kernel. Consequently, $u_M(\xi_D) = \exp(-|\xi_D|/\gamma)$ should be valid for such experiments, by taking into account that its Fourier transform is just a Lorentzian function (Gaskill, 1978). The parameter γ specifies the size of the uncertainty region around the real point emitter that produces the

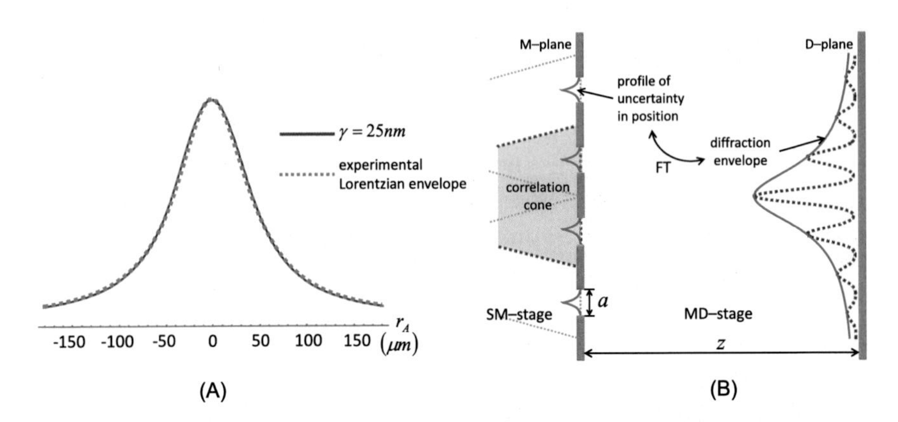

Figure 12 (A) Comparison of the experimental diffraction envelope in Fig. 10 with the diffraction envelope calculated from Eq. (15) with $u_M(\xi_D) = \exp(-|\xi_D|/\gamma)$ and the following entries: $\lambda = 4.6$ pm, $z = 1.25$ m, $a = 50$ nm. Each slit contains 11 real point emitters with spacing 5 nm. The optimal fit is obtained for $\gamma = 25$ nm. (B) Conceptual sketch describing the relationship between $u(\xi_D)$ in each individual slit and the diffraction envelope of the interference pattern. FT means Fourier Transform.

Lorentzian diffraction modulation of the experimental interference pattern. In this sense, $L_C = 2\gamma$.

Fig. 12(A) shows the best fit between the profiles of the Lorentzian modulation of the experimental interference pattern with single C_{60} molecules and the Lorentzian diffraction envelope obtained from Eq. (15) and the experimental parameters, i.e. $\lambda = 4.6$ pm, $z = 1.25$ m, $a = 50$ nm. According to this result, $L_C = 2\gamma = 50$ nm, i.e. the region of uncertainty in position of the C_{60} molecules covers the individual slit so that the molecules going through any point in the slit can be represented by a single real point emitter, placed at the slit midpoint, with uncertainty region equal to the slit width. This description is conceptually depicted in Fig. 12(B). It also implies that, in an interference experiment, single molecules behave as relatively localized objects. Thus, the assumption of "particle delocalization" and the notion of wave-collapse are not necessary for explaining particle diffraction. Finally, it is worth noting that, in case of molecules, the angular spread of the Lorentzian cones is narrower than that in case of light waves. For instance, the profiles in Fig. 6(B) depict Lorentzian cones for light waves at a distance of 6.32 cm from the emitting pinhole, while the Lorentzian envelopes of the interference patterns in Fig. 11 were recorded at a distance of 1.25 m. These results show that the uncertainty in position of molecules is larger than that of light waves.

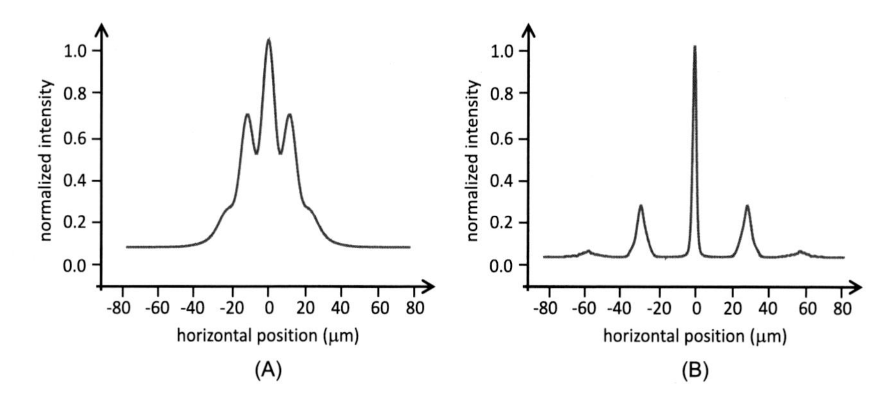

Figure 13 Theoretical profiles of the interference patterns in experiments with single molecules of (A) $F_{24}PcH_2$ and (B) PcH_2 reported in Juffmann et al. (2012). The different contrasts of the profiles are obtained by designing appropriately the SM-preparation stage.

The detailed analysis reported in Castañeda et al. (2016a), regarding the interference experiments with PcH_2 and $F_{24}PcH_2$ molecules (Juffmann et al., 2012), confirms the validity of the new interpretation. Fig. 13 shows the profiles of the interference patterns calculated with this methodology. Their excellent fit with the experimental results is detailed in Castañeda et al. (2016a) and not included here. In the framework of the present model, we wish to point out that the influence of the van der Waals interaction between the molecules going through the slits and the grating material as well as the molecular fractioning or diffusion on the detection substrate are not needed to describe accurately the observed interference patterns reported by Juffmann et al. (2012).

The analysis above reinforces the requirement of discreteness of the set of real point emitters (i.e. the zeroth order class) in order to produce diffraction modulation. However, the condition for wave diffraction established in Castañeda & Muñoz (2016), Castañeda (2017a) should be generalized as $L_C < b \leq 10L_C < L$, thus including the uncertainty in position for waves or particles. This generalized criterion plays an important role to evaluate accurately stringent requirements of the experimental design.

7. CONCLUSION

The main features of the new unified model to account for interference of waves and material particles have been reviewed in Section 4. We have demonstrated that the two-point correlation properties of the space enclosed by the experimental set-up in combination with particle momentum provide a careful description of interference of light or of material particles without using a wave-optical model. As a consequence the controversial hypotheses regarding the interpretation of the wave probability function are removed.

Suffice it to say that, at present, we are not able to explain the fundamental nature of the spatial correlation, but we have reported a physical intuitive and rigorous theory that one can calculate with to obtain striking results. The (non-local) interactions, between each real point emitter and the virtual point emitters, that can be considered as the heart of the *new interference principle*, emphasize a peculiar role of the interferometer configuration and the space it encloses. We hope that these considerations will help to shed new light on the role of space in a basic problem of quantum mechanics.

Finally, preliminary considerations are reported about the importance of the role of the uncertainty principle to explain the intensity modulation (diffraction effect) of interference fringes of massive molecules. We showed that diffraction curves are calculated precisely by considering that, at an aperture plane, a line of real point emitters, with a given spacing, can be replaced by a single point emitter located at the midpoint of the array. It means that all the point emitters, within a distance from the midpoint equal to half of the array length, are not distinguishable so that such distance is assumed as the uncertainty in position of particles passing through the point P. A more detailed work is in progress along this line.

ACKNOWLEDGMENT

We are deeply grateful to Peter W. Hawkes for his ongoing support to disseminate our unconventional ideas.

REFERENCES

Arndt, M., Nairz, O., Vos-Andreae, J., Keller, C., van der Zouw, G., & Zeilinger, A. (1999). Wave–particle duality of C_{60} molecules. *Nature*, *401*, 680–682.

Bach, R., Pope, D., Liou, S. H., & Batelaan, H. (2013). Controlled double-slit electron diffraction. *New Journal of Physics*, *15*, 033018, 7 pp.

Born, M., & Wolf, E. (1993). *Principles of optics* (6th ed.). Oxford: Pergamon Press.

Castañeda, R. (2014). Electromagnetic wave fields in the microdiffraction domain. *Physical Review A*, *89*, 013843, 14 pp.

Castañeda, R. (2017a). Discreteness of the real point emitters as a physical condition for diffraction. *Journal of the Optical Society of America A*, *34*, 184–192.

Castañeda, R. (2017b). Interaction description of light propagation. *Journal of the Optical Society of America A*, *34*, 1035–1044. The concept of geometric potential was firstly introduced for wave interference in this work.

Castañeda, R., Matteucci, G., & Capelli, R. (2016a). Quantum interference without wave–particle duality. *Journal of Modern Physics*, 7, 375–389.

Castañeda, R., Matteucci, G., & Capelli, R. (2016b). Interference of light and of material particles: A departure from the superposition principle. In P. W. Hawkes (Ed.), *Advances in imaging and electron physics, vol. 197* (pp. 1–43). Burlington: Academic Press.

Castañeda, R., & Muñoz, H. (2016). Spectrum of classes of point emitters: New tool for nonparaxial optical field modeling. *Journal of the Optical Society of America A*, *33*, 1421–1429. The spectrum of classes was firstly introduced for wave interference and diffraction in this work.

Feynman, R. P., Leighton, R. B., & Sands, M. (1965). *The Feynman lectures on physics, vol. 3.* Menlo Park, CA: Addison-Wesley.

Frabboni, S., Gabrielli, A., Gazzadi, G. C., Giorgi, F., Matteucci, G., Pozzi, G., ... Zoccoli, A. (2012). The Young–Feynman two-slit experiment with single electrons: Build-up of the interference pattern and arrival-time distribution using a fast-readout pixel detector. *Ultramicroscopy*, *116*, 73–76.

Gaskill, J. (1978). *Linear systems, Fourier transforms and optics*. New York: John Wiley & Sons.

Juffmann, T., Milic, A., Muellneritsch, M., Asenbaum, P., Tsukernik, A., Tuexen, J., ... Arndt, M. (2012). Real-time single-molecule imaging of quantum interference. *Nature Nanotechnology*, 7, 297–300.

Juffmann, T., Truppe, S., Geyer, P., Major, A. G., Deachapunya, S., Ulbricht, H., & Arndt, M. (2009). Wave and particle in molecular interference lithography. *Physical Review Letters*, *103*, 263601.

Kocsis, S., Braverman, B., Ravets, S., Stevens, M. J., Mirin, R. P., Krister Shalm, L., & Steinberg, A. M. (2011). Observing the average trajectories of single photons in a two-slit interferometer. *Science*, *332*, 1170–1173.

Mahler, D. H., Rozema, L., Fisher, K., Vermeyden, L., Resch, K. J., Wiseman, H. M., & Steinberg, A. (2016). Experimental nonlocal and surreal Bohmian trajectories. *Science Advances*, *2*, e1501466.

Mandel, L., & Wolf, E. (1995). *Optical coherence and quantum optics*. Cambridge: Cambridge University Press.

Matteucci, G. (2011). On the presentation of wave phenomena of electrons with the Young–Feynman experiment. *European Journal of Physics*, *32*, 733–738.

Matteucci, G. (2013). Interference with electrons – from thought to real experiments. *Proceedings of SPIE*, *8785*, 8785CF.

Matteucci, G., Pezzi, M., Pozzi, G., Alberghi, G., Giorgi, F., Gabrielli, A., ... Gazzadi, G. (2013). Build-up of interference patterns with single electrons. *European Journal of Physics*, *34*, 511–517.

Nairz, O., Arndt, M., & Zeilinger, A. (2003). Quantum interference experiments with large molecules. *American Journal of Physics*, *71*, 319–325.

Zeilinger, A., Gaehler, R., Shull, C. G., Treimer, W., & Mampe, W. (1988). Single and double-slit diffraction of neutrons. *Review of Modern Physics*, *60*, 1067–1073.

A Review of Scanning Electron Microscopy in Near Field Emission Mode

Taryl L. Kirk
Rowan University, Glassboro, NJ, USA
Educational Testing Service, Princeton, NJ, USA
e-mail address: kirk@rowan.edu

Contents

Advances in Imaging and Electron Physics, Volume 204
ISSN 1076-5670
https://doi.org/10.1016/bs.aiep.2017.09.002

1. INTRODUCTION

For many years there has been an increasing trend towards using scanning electron microscopy (SEM) with lower beam energies and this tendency is expected to continue. In low voltage SEM (LVSEM), the penetration depth of the impinging electrons is small, which gives rise to greater surface sensitivity. Consequently, the penetration depth of the impinging electron beam reduces towards the escape depth – nullifying the so-called "edge effect." In addition, the secondary electron (SE) yield is higher and the total emitted signal approaches unity, which also reduces charging in semiconducting and insulating samples. Novel LVSEM techniques, *e.g.* Very Low-Energy SEM (VLSEM), allow for crystalline, diffraction, and dopant contrast mechanisms (see Müllerová & Frank, 2007). Although VLSEM delivers surface sensitivity with numerous contrast capabilities, it does not exhibit the high resolution observed with the scanning probe microscopies, (SPMs), such as scanning tunneling microscopy (STM). In STM, the interaction occurs between the electrons in the orbitals of the tip apex and the sample surface. When combined with the ultrahigh-precision position resolution of the piezoelectric device – of the order of picometers – used to maneuver the tip, atomic structures can be observed (Binnig & Rohrer, 1982).

In this chapter, we report on the combination of the aforementioned types of microscopy into a single technique "Near Field Emission Scanning Electron Microscopy" (NFESEM) (Kirk, Ramsperger, & Pescia, 2009) that combines some of the best features of VLSEM and STM. In essence, NFESEM is an intermediate technique in which electrons are emitted from a needle tip via field electron emission (FE), and then impinge on and interact with the sample. As a result, electrons are ejected from the sample surface and detected. NFESEM differs from VLSEM, in that there is no remote electron gun column. Instead, the electron source is positioned locally using a four-quadrant piezoelectric tube, commonly used in STM. However the field emitter is positioned at a distance much further from the sample than in STM.

Moreover, because the voltage V_a applied between the tip and the sample has a magnitude much greater than the sample work-function, the total energy of the field emitted electrons is also well above the sample vacuum level; therefore, the incoming electrons enter the sample with energies much higher than in conventional STM. The basic components of the NFESEM will be described in detail in Section 2. Moreover, this section will also discuss how measurements are made and how the NFESEM differs from similar localized field emission mode microscopy techniques (see Section 2.4).

Another subtle difference is that the probe geometry is usually unknown and irrelevant in STM and other scanning probe microscopies (SPMs). In the SPMs only one apex atom (or a cluster of apex atoms) interacts with the sample, whatever the mesoscopic shape and structure of the probe. In contrast, NFESEM requires that the electron source be well defined and symmetric – as described in Section 3 – since the FE current is dependent not only on the apex structure but also on the overall tip shape (further detail in Section 4.2) and on its chemical nature (see also Kirk, De Pietro, Pescia, & Ramsperger, 2009). We will demonstrate how the needle tip formation can be controlled; most importantly, we will correlate the geometry of the emitter with its electrostatic behavior (see Sections 3.2 and 4.4). This can potentially be a powerful tool not only for NFESEM research, but for SPM users in general. We have worked extensively on optimizing this process, which has become a central theme in our studies.

Conventional SEM offers high-resolution imaging capabilities, almost independent of the electron-source shape and dimensions, due to the use of electro/magneto-static lenses. These lenses also introduce aberrations that limit the spot size of the impinging electron beam. Although it is possible to reduce the electron gun column to microscopic dimensions – accordingly reducing the aberration effects – or to introduce instrumental or simulation-based aberration corrections, the aberrations will still be present. NFESEM technology circumvents this issue by eliminating all of the electro/magneto-static lenses, and using the sample itself as the extracting electrode. NFESEM does not employ lenses, but its geometry allows for a self-focusing (*i.e.*, "beam-concentrating") effect (Kyritsakis, Kokkorakis, Xanthakis, Kirk, & Pescia, 2010). This effect is characteristic of FE from needle-shaped emitters with a radius of curvature, r_{tip}, of the order of the barrier width. Field emitters of this class do not follow conventional Fowler–Nordheim (F–N) theory (discussed in Section 4.2), but instead exhibit non-planar FE (see He, Cutler, & Miskovsky, 1991; Cutler, He, Miskowsky, Sullivan, & Weiss, 1993). Section 4.3 will examine how this self-focusing effect enhances the lateral resolution capabilities of NFESEM.

As NFESEM positions the field emitter using piezoelectric tubes similar to STM, the same scanning modes are available, *i.e.* a constant height (CH) and a constant current (CC) mode. In CC mode the field emitted current is kept constant. Both modes are advantageous for specific measurements, but are distinct from STM because the NFESEM operates at larger distances – tens of nanometers. For instance, the "approach," or initial positioning of the tip in tunneling contact (< 1 nm) with the sample surface, is performed in CC mode. It is also important to note that if the emission current is held fixed, the local electric field at the tip surface is also fixed (or very nearly so). This will be explored further in Section 4.4.

Contrary to contemporary SPMs, NFESEM does not generate images via localized interaction between probe and sample; rather, images are generated via the influence of tip-sample distance on the probe current. This allows NFESEM to generate images in either CC or CH mode (see Section 2.1). Incidentally, SEM is also operated in CC mode, *i.e.* the primary beam current does not vary with the surface topography as it is raster-scanned over the specimen. The secondary-electron (SE) yield – in SEM – is strongly dependent on the local curvature of the sample surface as well as its chemical composition (see Goldstein et al., 2003). One of the main features of NFESEM is that imaging can be performed in CH mode; in this mode the variations in the FE current can be used to enhance the SE yield (also discussed in Kirk 2010a, 2010b). Section 3.2 will highlight how this feature can be used to achieve atomic vertical resolution, a capability that eluded early NFESEM-type devices operating in CC mode (see Section 2.4).

Contrast mechanisms in SEM are not limited to subsequent analysis of the SEs; for example, the beam-sample interaction is strongly dependent on the energy of the impinging electrons (there are additional effects discussed in Cazaux, 2010). This requires knowledge of how the SE yield varies with the energy of the primary electron beam. A contrast inversion occurs when the measured SE current exceeds the primary electron beam current. This inversion is material dependent, and often occurs at energies ranging from 100–1000 eV. The lower end of this range is within the operating range of the NFESEM. Therefore, it is imperative that the SE detector (SED) be calibrated (see Appendix A), in order to determine the SE current impinging on the detector. CH mode provides a topographic template of the surface that can be "subtracted out" to yield purely SE-based characteristics, comparable to conventional SEM. We will present how the topographic images are analyzed in Section 5, which describes the origin of the detected electrons. This is essential when investigating surface properties that are not of a topographic nature, *e.g.* surface magnetization, chemical composition, doping concentration, and other alternative contrast modes (some of these options are explored in Section 6).

1.1 Motivation and Terminology

NFESEM is not intended to replace VLSEM nor STM; but rather it offers an alternative technique to users who want to perform measurements on "clean" conducting surfaces that have a corrugation depth less than half of the tip-sample separation gap. NFESEM was originally intended to increase the resolution capabilities of a SEM-based instrument. We emphasize that the CH *modus operandi* is the major advance in our technology, because it allows for SEM with extreme vertical resolution. Furthermore, NFESEM introduces a method to generate two complementary images (FE- and SE-based), increasing the user's investigative capabilities. Accurate calibration allows for SE-yield extraction, which may lead to additional contrast mechanisms when compared to the total emitted signal. On a similar note, the calibrated signal can be used to separate the surface topography from the SED-dependent signal. We expect a similar spatial resolution for both *modi operandi*, assuming equivalent extraction efficiencies.

Initially, the NFESEM was intended to be coupled with a Mott spin polarimeter, in order to perform SEM with polarization analysis (SEMPA) (Koike & Hayakawa, 1984). Currently, we are developing a stand-alone SEMPA system, with the aim of incorporating a spin polarimeter into a NFESEM instrument later. In ordinary SEMPA, a remote source produces unpolarized electrons: magnetic contrast results from analyzing the spin of the electrons ejected from the surface. The electrons are extracted from the sample using an extraction voltage typically ranging between 200 V and 600 V. After initial acceleration, the electrons are further accelerated to 50 kV and are then passed through a 90°-deflection, which acts as a low-pass electron energy filter. This is used to select SEs that had emission energies of 5 eV or less, since these show the highest polarization for the transition metal magnets (Koike & Hayakawa, 1984). These SEs are then accelerated towards a gold foil and undergo scattering processes. The spatial resolution of our current SEMPA system, which is based on a remote electron source, is $\sim$ 20 nm; this resolution is mainly restricted by the electron spot size, which limits the imaging capabilities of magnetic domain walls. We plan to develop a new form of electron source, because high resolution electron gun columns are very costly and not easily suited to ultra-high vacuum (UHV) conditions. Magnetic-contrast imaging using an NFESEM-type configuration with a spin polarimeter has already been performed in CC-mode (see Allenspach & Bischof, 1989; First, Stroscio, Pierce, Dragoset, & Celotta, 1991). Although these results were promising, research activities along this direction have been abandoned.

An alternative approach is to excite a magnetic sample with a spin-polarized beam of electrons, similar to spin-resolved inverse photoemission (IPE) (Passek, Donath, & Ertl, 1996) and spin-polarized low energy electron microscopy (SPLEEM) (see Bauer, Duden, & Zdyb, 2002). This results in variation of the SE yield due to spin-dependent interaction during electron scattering (see Section 6.1). We believe that advances in these types of magnetic-contrast imaging mode will yield high spatial resolution capabilities.

Scientific communities associated with a specific technique often develop terminology of their own, in order to simplify discussion as well as stress the important characteristic parameters. Although NFESEM is in its infancy, it is no different. For instance, the term "near" refers to the location of the electron source, as compared to conventional remote-electron-gun sources. This should not be confused with near-field scanning optical microscopy (more commonly known as NSOM/SNOM), which is a SPM technique that employs optical illumination of the specimen. NFESEM is a SEM technique that, by placing the field electron emitter "near" the sample under investigation, incorporates the high positional resolution available in SPMs. Note that, within the technology developed in our laboratory, conventional STM can also be performed prior (or subsequent) to NFESEM measurements.

The total current entering the detector contains some back-scattered primary-beam electrons (BSEs) as well as true secondary electrons (SEs) generated in the sample surface layer. However, the proportion of the SED output signal that relates to BSEs is relatively small, so – in accordance with common practice in the literature – we call the total current I_{SE} entering the detector a SE current, and treat the detector as a SE detector.

Typically, the SE signal I_{SE} is determined as the product of the following quantities: the probe current I_P provided by the field emitter, the SE yield δ, and the collection efficiency f_{SE}. The product $f_{SE}\delta$ is called the "effective SE yield efficiency", and can be determined after calibrating the detector, as described in Appendix A. For qualitative analysis, we will refer to the total emitted signal from the sample; for quantitative analyses we will use the effective SE yield efficiency.

Another important distinction is our description of the beam energy. For short tip-sample separation distances, as used in STM and in a STM-like regime in NFESEM, the electrons tunnel directly from the tip to the sample. But, in NFESEM, the distance d between the tip and the sample surface is much larger than the barrier width L; in this case, the electrons are physically emitted from the source into the vacuum and are then accelerated by the extracting field. On reaching the sample, they have (before entry) a kinetic energy approximately equal to $(eV_a - \varphi_s)$, where φ_s is the sample work-function. There is a small energy spread, typically a fraction of an eV, originating from the details of the emission process.

The important measurables of the NFESEM are as follows:

- $\boldsymbol{I_P}$ – the probe current, *i.e.*, emission current emitted by the tip, as measured by the current flow to the tip; this is determined by the field distribution at the tip surface, and this in turn is determined by the applied voltage V_a and the geometry of the system, including the topography of the sample surface; note that the probe current is represented by the symbol I_{FE} in the FE theory described later;
- $\boldsymbol{I_{SE}}$ – the "SE signal", *i.e.* the electron current captured by the detector;
- $\boldsymbol{f_{SE}\delta}$ – the effective SE yield efficiency, determined as the ratio I_{SE}/I_P;
- $\boldsymbol{E_P}$ – the probe energy; this is fixed during imaging and is defined as equal to eV_a, where (as above) V_a is the magnitude of the voltage difference applied between the tip and the sample (note: V_a rather E_P will be used in FE theory);
- $\boldsymbol{d}$ – the tip-sample separation, *i.e.*, the gap between the tip and the sample. For working purposes, d is determined as the distance by which the tip is retracted from its position when it is emitting in the STM-like tunneling regime. The vertical distance scale is calibrated using known features of the surface (*e.g.*, monoatomic steps), as measured by lateral scanning in the STM-like regime;
- $\boldsymbol{r_{tip}}$ – the apex radius of the field emitter tip. This is the tip's physical radius, as modeled by (for example) fitting a sphere to the apex of a TEM or SEM image of the tip. Estimates of "effective tip radius" can be determined, in various ways, using F–N theory (see Sections 4.1.1 and 4.2);
- $\boldsymbol{\Delta x}$ – the experimental lateral resolution; this is primarily determined by the size of the beam spot at the sample surface, and is determined experimentally by measuring the point spread function resulting from scanning the beam across an atomically sharp edge;
- ***LR*** – the theoretical lateral resolution, simulated by determining the trajectory of the electrons emitted by the source in accordance with the system geometry (especially the tip geometry);

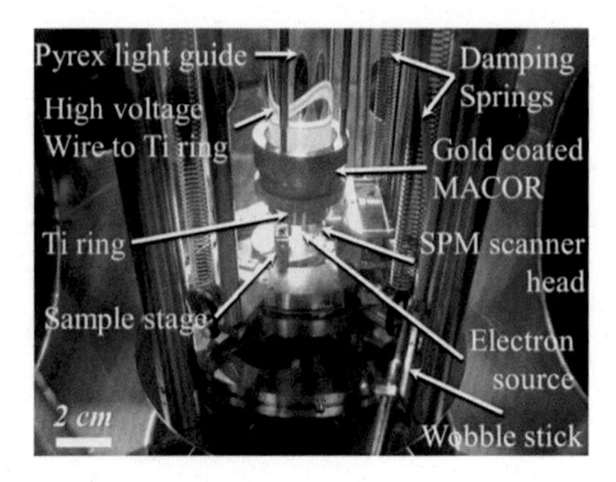

Figure 1 First NFESEM prototype (Kirk, 2010a).

- ***L*** – the barrier width; this is given approximately by $L \approx \varphi/eF$, but exact values need to take into account both image-potential energies and (for nanometer-scale field emitters, say $r_{\text{tip}} < 20$ nm) the tip curvature.

Other standard SPM parameters include scan area and speed, but have not yet been investigated in detail.

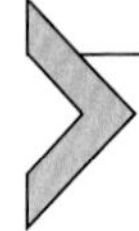

2. INSTRUMENTATION

2.1 First Prototype

The present NFESEM system, see Fig. 1, consists of a homemade modified Lyding-type scanning tunneling microscope and a SED. UHV conditions are required to reduce surface contamination, which significantly alters the SE yield; additionally, this environment increases the primary beam stability. Therefore, the system was installed in a specially designed titanium UHV chamber with an inner aluminum coating, enabling the system to achieve a base pressure as low as $2 \cdot 10^{-11}$ mbar. Moreover, the STM is suspended by four stainless steel springs for high frequency damping and additionally has eddy current damping on the bottom side of the base plate. The geometry of the STM allows for the detector to be mounted near the sample. Our SED is situated approximately 2 cm from the sample edge and aligned to collect electrons ejected parallel to the surface; where the highest SE signal was experimentally observed.

The probe current is typically of the order of tens of nano-amperes. Higher currents may increase the vertical resolution of the image (see Section 3.2), but they may also contaminate the surface and/or induce adsorbate transfer from the sample to the tip. Suitable experimental conditions have to be found by trial and error. The probe energy E_P should be the minimum energy required to eject several electrons from the top-most layer of the surface, and is usually less than 60 eV. SEs and BSEs will be deflected by the strong electric field ($\sim$ 1–3 V/nm) between the tip and the sample; this field repels the escaped electrons toward a direction running parallel to the surface. The electron detector, mounted with its axis essentially parallel to the sample surface, should cover an area large enough to encompass the entire sample. An acceleration voltage may be applied to the scintillator disk or extraction optics; however, it should not perturb the primary beam. Fig. 1 shows a photo-multiplier tube (PMT) and a scintillator used for detecting the SEs. The extraction voltage used is determined by the minimum acceleration voltage required to excite measurable photons at the scintillator disk, located at the entrance.

The SEs are accelerated to a $YAlO_3$ single crystal scintillator (ø = 2 cm), in the perovskite phase, (YAP) at energies up to 3 keV; here they are converted to photons that travel along a 45 cm Pyrex light guide, as depicted in Fig. 1, to a Hamamatsu R268 PMT. A gold-coated MACOR piece insulates the high voltage at the titanium ring and is at the same ground as the detector. Our detector, comprised of the scintillator and PMT, is mounted on a linear UHV z-manipulator with a linear travel distance of 10 cm. The PMT is then mounted on the outer (or ambient) side of a quartz window flange, which enables the conversion of the signal from photons back into an electrical signal.

In general, the probe energy E_P is on the order of tens of eVs; hence the probe beam generates reflected or ("back-scattered") electrons (BSEs). Both BSEs and SEs are sampled in the present setup, due to a large detector surface area (3.14 cm^2) and high electron post-acceleration voltages. Typically, the detector angle of acceptance is 45°, as the *modus operandi* places the detector within two centimeters to the center of the sample. This SED is reminiscent of an Everhart and Thornley (ET)-type detector, with the exception that the Faraday cage precedes the scintillator. Both BSEs and the SEs have low energies, due to the low probe energy. For this reason, instead of using an accelerating grid, the electrons are accelerated directly toward the scintillator, which is surrounded by a conducting titanium ring at high voltage ($U_{\text{Ti-ring}} \geq 1.75$ kV). In general, the ET detectors are most sensitive to low-energy SEs, since the solid angle of collection is too small for BSEs.

2.2 Alignment and Operation

Measurements can begin immediately after the field emitter and the sample have been transferred from the preparation chamber to the measurement chamber. The measurement procedure has four stages:

i) manual alignment of the sample stage;
ii) tip approach using coarse (or long-range) piezo motors;
iii) automatic tip approach using the Nanonis program (Nanonis is now SPECS Zürich GmbH);
iv) measurement, involving one or more of
- **a)** NFESEM imaging,
- **b)** optional STM reference measurement,
- **c)** alternative electrostatic characterization.

The Digital Signal Processor (DSP) card (see Kirk, 2010a) is switched on first, allowing the computer to control and partially execute the procedure. This card will control, in real time, every feedback from/to the system. Furthermore, the Nanonis program and a piezo remote-controller on the computer are also switched on, in order to control and prevent discharges. The sample stage can now be aligned to center the tip on the sample surface. The main task of the alignment procedure is to simplify and accelerate the next procedural step, the manual approach of the field emitter.

The piezo controller has to be calibrated by changing the frequency and the voltage applied to the piezo mover before initiating a tip approach, since the constants that define the mobility of the piezos are strongly dependent on the system conditions. Once the sample stage and the field emitter both move properly inside the chamber, the sample stage is mechanically floated, isolating the measurement area from high-frequency vibrations; a manual tip approach is then executed. This is done by manually approaching the sample to the field emitter, with the aid of a monocular. The table, on which the measurement and preparation chamber are mounted, is mechanically floated when the distance between the field emitter and the sample is within a millimeter. At this point, the measurement system is fully isolated from interference coming from its surroundings, and the automatic tip approach can be performed.

In the automatic tip approach, the gap between the tip and the sample is reduced until a set tunneling current – decided *a priori* – is reached, as determined automatically in CC mode by the Nanonis program. This set current is chosen in such a way that it occurs in the STM-like regime, when the tip is about 1 nm or less from the sample surface. This approach is executed by using the cyclic repetition of the following steps:

i) scan the tip towards the sample surface (*i.e.* in the negative z-direction), by a distance of up to 600 nm via a fine movement piezo tube scanner;

ii) then

  (1) if the set tunneling current is measured during this scan, then the scan stops and the approach procedure is considered completed;

  (2) if the set tunneling current is not measured, then:

    (a) retract the tip to the starting position,

    (b) move the sample towards the tip by approximately 300 nm with a coarse movement piezo tube scanner,

    (c) repeat the cycle, starting at (i).

The scan and retraction are done using the fine motion of a small "z-motion" four-quadrant piezo tube on top of the coarse x–y piezo scanners.

The automatic approach is terminated using well-established values for the probe current and applied voltage; namely, $I_P = 0.15$ nA and $V_a = 200$ mV (for metals). These values, characteristic of the STM-like tunneling regime are used to define a "zero-position"; motion inside the zero position has to be avoided, in order to prevent mechanical contact between the tip and the sample, with probable damage to both.

After this first automatic tip approach, the field emitter is sometimes retracted slightly, and an additional automatic approach is performed in order to place the tip in a more suitable position along the z-axis. If necessary, the tip is retracted and repositioned laterally, in order to avoid surface debris encountered during the first approach. This is a standard STM procedure.

2.3 NFESEM Imaging

Upon the retraction of the tip to the pre-set scan height, the voltage between the tip and the sample is increased until the desired field emission (probe) current is achieved. If the tip radius is large, higher voltages will be needed to generate a given probe current. We find that tips with the smallest attainable radii usually give the best vertical and lateral spatial resolution. However, tip shape stability may be problematic in high electric fields and currents; this is a drawback of using very sharp tips. Clearly, the routine experimental performance will be the result of compromising between various requirements.

In order to generate an image, the tip is rastered along the sample surface with the z-motion servo switched off, *i.e.* CH mode is used, which is obtained using a sample-and-hold amplifier (please refer to the electronic block diagram of the NFESEM in Kirk, 2010a). Tip-sample orientation is another crucial factor in performing NFESEM measurements, as the scans should be made parallel to "flat" surfaces. Small FE probe currents were employed to correct for the tip alignment by minimizing the current deviation in both the *x*- and *y*-directions at several heights, before initiating a full scan at the desired height. The deviations in probe current are very sensitive to its magnitude; therefore we have performed the alignment with currents of 50 pA to 300 pA. FE instabilities, resulting in current spikes and steps, significantly distort the signal, thus inhibiting alignment. Careful tip and sample preparation, aimed at removing all oxides and other contamination, is the only method to prevent such instabilities.

We have additionally scanned the tip at distances between 100 nm and 120 nm to the sample surface, using biases up to 60 V. This field-shaping technique often forms an asperity smaller than 5 nm on the tip apex; its formation is characterized by a large increase in FE current at a fixed voltage and height. Finally, the tip height is reduced to 20–60 nm[1] for high resolution NFESEM imaging. Additional tip sharpening can also occur at these heights or during prolonged scanning.

2.4 Related Field Electron Emission Scanning Microscopy Techniques

This section describes similar local field emission electron scanning electron microscopy techniques implemented prior to the NFESEM. R.D. Young et al. were the first to engineer a localized field emitter tip as a means of generating a primary electron beam (see Young, Ward, & Scire, 1972). In the experimental set-up proposed by R.D. Young, see Fig. 2, a tip was brought into close proximity to the sample surface until the desired FE current is achieved. The tip was then rastered along the surface at a fixed FE current, *i.e.* CC mode, which was maintained by a feedback controller that adjusts the vertical *z*-piezo voltage accordingly. The applied voltage was much larger than the work function of the emitting material, which results in a field emission current–voltage (I–V) relation consistent with conventional Fowler–Nordheim (F–N) field emission (Fowler & Nordheim, 1928). This mono-energetic primary beam of electrons impinged on the surface at a well-defined area, without the aid of focusing optics, and underwent elastic as well as inelastic collisions penetrating the outer-most

[1] Lower scan heights, less than 10 nm, were used while imaging with the second NFESEM prototype (see Section 6.1).

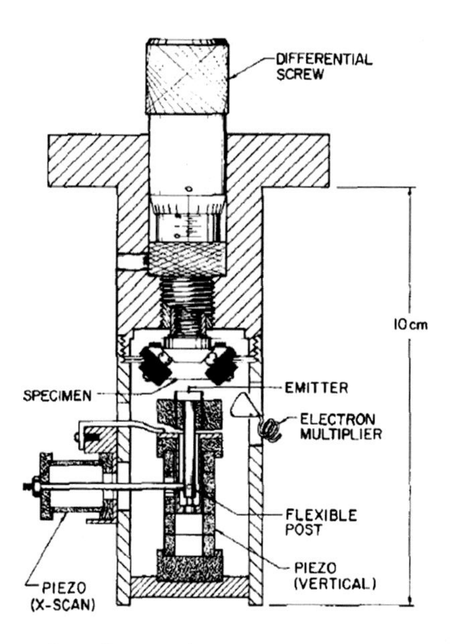

Figure 2 Original topografiner design. This instrument was the predecessor of the scanning tunneling microscope; however, the image was generated via the modulations of the secondary electron signal (Young et al., 1972). *Reprinted with permission from AIP Publishing LLC.*

layer(s) of the surface (less than 1 nm). Subsequently, the ejected SEs from the excitations were collected and detected using an electron multiplier. The variations in the electron intensity from the electron multiplier signal, as a function of scanning position, represented the surface topography. Although this experiment by R.D. Young et al. is conceptually similar to a STM, however it generates images comparable to a SEM without the need of a "remote" electron gun.

More than a decade later, H.-W. Fink achieved 3 nm lateral resolution on a polycrystalline gold surface by measuring the electron intensity from an electron multiplier (see Fink, 1986), resulting from the interaction of a field emitted beam of electrons from a single atom (111)-oriented tungsten (W)-tip. Subsequently, the primary beam of electrons was extracted from the STM tip via an aperture placed in between the tip and the sample. These single-atom-apex tips require field evaporation, *e.g.* with field ion microscopy (FIM), to atomically engineer the apex. The tip was rastered near the surface using CC mode at an emission current of 0.1 nA and primary beam energy of 15 eV. In addition, the primary electron beam diameter was reduced by the aperture. Notably, the inclusion of apertures limits spacing, and prevents the positioning of the tip closer to the sample. This was a major advancement; however, single crystal *W* wires are expensive and these single atom tips require a field ion microscope to sharpen the apex.

2.4.1 Electron Spin Polarization

The NFESEM is a very powerful tool for the investigation of topographic nanometer-sized features on level surfaces (see Kirk, 2010a); additionally this same electron microscopy technique can be used to observe other contrast mechanisms, as the SEs also contain magnetic and chemical information about the sample. In particular, the domain formation mechanisms of the two-dimensional surface magnetization vector can be studied by employing the polarization analysis of low energy SEs generated via scanning electron microscopy; more specifically, SEMPA. K. Koike et al. devised this instrument to improve the imaging of magnetic material using SEM (Koike & Hayakawa, 1984); that previously involved (1) the deflection of electrons ejected from the surface in an external magnetic field (Joy & Jakubovics, 1968), and the other concerns (2) the deflection of reflected electrons in a magnetic field within the magnetic material (Philibert & Trixier, 1969). The SEM can therefore be sensitive to the magnetic properties of a sample via the interaction of an unpolarized, primary electron beam with a magnetized sample. The ever decreasing size of devices based on the magnetic properties of materials has led to the need to image the magnetic domains or even spin configurations within the domain wall with high spatial resolution. In fact, much of the studies of systems exhibiting giant magneto-resistance (GMR), which led to the 2007 Nobel Prize in Physics and greatly enhanced the storage capacity of memory devices, were conducted using SEMPA. Similarly, the NFESEM can be upgraded for polarization analysis of the SEs, *i.e.* NFESEMPA, which will provide even higher resolution than the SEMPA, since the critical parameter limiting resolution is the impinging beam width (see Allenspach, 1994). The field emitter can also be replaced by a spin-polarized source, where the field emitted electrons exhibit a preferential direction of the electron spin, *i.e.* spin-polarized NFESEM (SNFESEM). Here the magnetic contrast originates from the variations in the detected SE yield.

Both methods were postulated and considered by D. Pierce in his discussion on *Spin-Polarized Electron Microscopy* (see Pierce, 1988), section three and four respectively. In his account of the feasibility of SNFESEM, he confers that there is a variation in scattering intensity on ferromagnetic samples due to a spin-dependent scattering asymmetry (SSA). It is estimated that this intensity is greatest for polarized electron beam energies of 100 eV or less, which is within typical NFESEM operating parameters. However, there is still one prevalent concern: the ability to distinguish between the topography and SSA. Sample roughness can greatly affect the SSA, because it depends on the angle of the incident beam with respect to the crystal planes. Although the SSA is typically smaller than the topographic signal, one can subtract the FE current signal from the SE signal to reveal pure SE features. In regards to NFESEMPA, it has been shown that high resolution images of the surface can be attained by measuring the electron intensity of byproducts expelled from the sample surface, *i.e.* both SE and backscattered electrons. A spin detector will be designed and constructed in order to perform polarization analysis of these electrons, hoping to achieve the high resolution using spin asymmetry as a contrast mechanism.

R. Allenspach was the first to implement NFESEM in magnetic contrast imaging mode; where he replaced a "remote" SEM electron gun with a STM to reduce the beam spot size (see Allenspach & Bischof, 1989). The energy spectrum of the electrons ejected from the sample exhibited a spectral behavior similar to conventional SEM. Accordingly, there was a large peak of conduction electrons ejected from the sample after undergoing multiple collisions during the cascade process, *i.e.* SEs. Spin polarization, which was recorded via a Mott spin polarimeter, peaks at the lowest detectable energies of the SEs. This pioneering experiment conducted by R. Allenspach at IBM, measured a hysteresis loop with secondary electrons emitted by a STM tip FE-excitation.

P.N. First et al. also performed similar measurements as described by H.-W. Fink, however spin analyzers – shown in Fig. 3A – were employed instead of an electron multiplier (see First et al., 1991). Topography, electron intensity, and surface magnetization of the sample were simultaneously measured with a lateral resolution of about 50 nm. The tip was rastered using *CC* mode at a nominal distance of 100 nm to the sample dynamically adjusted by a correspondingly controlled z-piezo using the "topografiner mode". Voltages ranging from 32–45 V were applied between the tip and sample. P.N. First et al. performed measurements on an array of rectangular Permalloy bits deposited on a silicon (Si) substrate. The resulting topographic and electron intensity image exhibited the inverse contrast, *i.e.* protrusions in the topographic image appeared as depressions in the electron intensity image (see Fig. 3B). Here, the topographic image is generated by the *z*-axis feedback of the STM, the electron intensity is the total

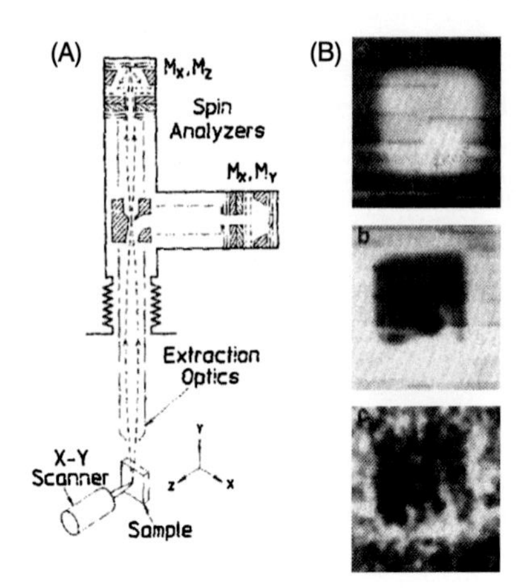

Figure 3 Magnetic contrast imaging using a spin detector. First et al. used a STM operating in FE mode to generate a primary beam of electrons that impinged and interacted with the sample (A). The subsequent electrons ejected from the surface were accelerated to a spin detector. (B) Topographic (top), detected electron intensity (middle), and spin asymmetry for magnetization along the *x*-direction of a sample consisting of an array of 50 nm thick rectangular Permalloy bits deposited on a Si substrate. The imaging parameters are the following: (scan area −1.12 μm by 1.12 μm) $d = 130$ nm, $I_{FE} = 10$ nA and $E_p = 38$ eV (First et al., 1991). *Reprinted with permission from AIP Publishing LLC.*

of all electrons detected (upon incidence with the sample), and the electron spin detector was positioned to measure spin asymmetries along the *x*- and *z*-directions. It is important to note that First et al. did not observe this contrast inversion when the same sample was imaged with a 10 keV SEM, *i.e.* SE yield was greater for the Permalloy bits *cf.* the Si. This is an important result, and contrast inversions will be discussed in Sections 6.1 and 6.2 in further detail.

2.4.2 Electron Energy Loss Spectroscopy

Alternatively R. Palmer et al. performed electron energy loss spectroscopy (EELS) using a single crystal (111)-oriented *W*-tip as a primary beam emitter (see Eves, Festy, Svensson, & Palmer, 2000; Festy, Svensson, Laitenberger, & Palmer, 2001). Here, an energy analyzer was used to determine chemical surface properties (device depicted in Fig. 4). The tip radius, as determined by SEM, was approximately 10–30 nm and obtained a lateral resolution of approximately 40 nm. Movement of the tip in the vertical direction was controlled by a feedback loop controlling the *z*-piezo, *i.e.* CC mode, to maintain a distance of less than 200 nm. The backscattered elec-

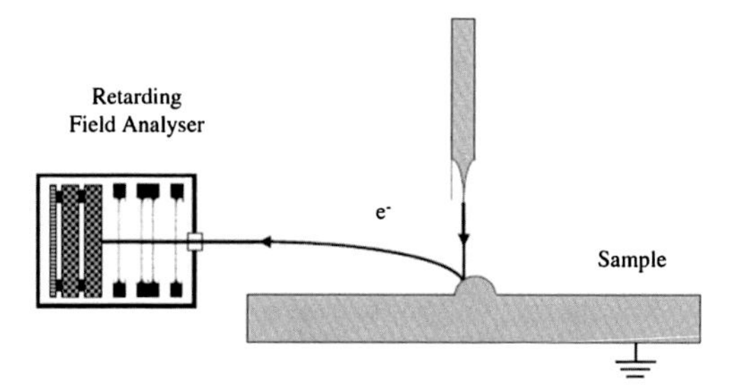

Figure 4 Sketch of the setup used to measure the energy of the electrons being ejected from the sample surface after be excited by electrons emitted from a STM tip (Festy et al., 2001).

trons entered the four-grid, retarding field analyzer through a grounded first grid. No direct relationship was drawn between the highest topographic area and the highest reflected electron count rate image; however angular resolved EELS was performed, where a distribution of BSEs and SEs was observed at angles ranging from −0.8° to 18°; with respect to the sample surface plane (Eves et al., 2000). The majority of electrons were ejected parallel to the surface, but electrons of the highest energy loss show trajectories bent towards the optical axis, in the direction of the primary beam. There are other techniques that involve SPM and various types of detectors, *e.g.* Auger electron spectrometer, and they are summarized in Wiesendanger (1994).

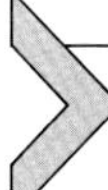

3. GEOMETRIC INFLUENCE ON FIELD EMISSION

3.1 The FE Process

Conventional F–N tunneling theory describes electric-field-induced tunneling through a roughly triangular – rounded, in practice – potential energy (PE) barrier. In the absence of an electric field, electrons are confined within the tip by the work function that separates the Fermi level from the local vacuum level. An applied local electric field F lowers the potential outside the emitter. If the field is strong enough, the probability of electron tunneling through the barrier becomes large, and a significant field electron emission current is emitted from states close to the emitter Fermi level.

In Fig. 5A, energies are measured relative to the bottom of the potential well. The y-axis denotes the energy component in the direction normal to the emitter surface. The "inner potential energy" (inner PE) χ is the total

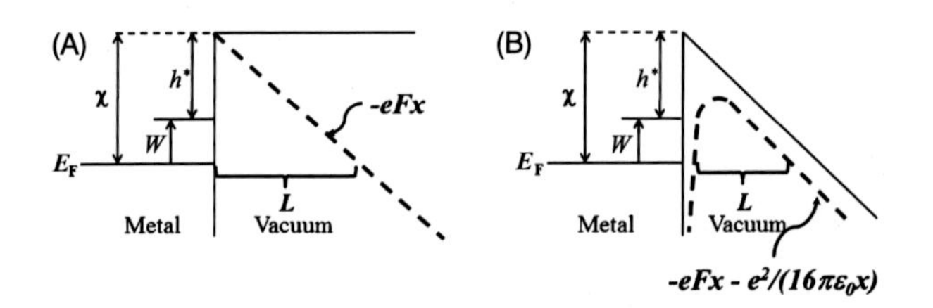

Figure 5 Schematic of the potential barrier altered by an external electrostatic field without (A) and with (B) the distortion due to the image potential (Kirk, 2010a).

height of the PE step encountered. W is the component of electron kinetic energy in the direction normal to the emitter surface (this is sometimes called the "forwards kinetic energy").

For an electron with forwards energy W, the exactly triangular PE barrier is defined by:

$$M(x) = (\chi - W) - eFx = h^* - eFx, \tag{3.1}$$

where the "zero-field barrier height" (h^*) is defined by:

$$h^* = \chi - W. \tag{3.2}$$

Although the Jeffreys, Wentzel, Kramers, and Brillouin (JWKB) semi-classical approximation can be applied to any barrier, F–N chose to solve the Schrödinger equation exactly for the exact triangular barrier depicted in Fig. 5A. The current density of the field-emitted electrons can be determined by evaluating the barrier penetration coefficient, $D(F, W)$, and then summing over all occupied electron states, by using Nordheim's supply function $N(W, T)$ for the number of electrons that strike the surface barrier per unit area per second with energy in the range dW. This yields the integral:

$$J(F, T) = e\int_0^{\infty} N(W, T)D(F, W)\,dW. \tag{3.3}$$

The so-called "elementary" F–N-type equation is based on the exact triangular barrier shown in Fig. 5A, and is a slightly simplified version of the F–N result. In modern form it can be written:

$$J = \left(a\varphi^{-1}F^2\right)\exp\left[-b\varphi^{3/2}/F\right]; \tag{3.4}$$

where the "first F–N constant" is given by $a = e^3/8\pi h_P \approx 1.541434 \cdot 10^{-6}$ eV V^{-2}, here h_P is Planck's constant, and the "second F–N constant" is noted as $b = (4/3)(2m_e)^{1/2}/e\hbar \approx 6.830890 \text{ eV}^{3/2}\,\text{V nm}^{-1}$. This equation, like all F–N-type equations, is a good approximation at room temperature.

Assuming an exact triangular barrier, as in Fig. 5A, allows a mathematically exact analysis, but the assumption is not physically realistic. A more reasonable model incorporates a classical image potential energy, which is depicted in Fig. 5B. For an electron with forwards energy W, this potential energy is described by:

$$M(x) = h^* - eFx - \frac{e^2}{16\pi\varepsilon_0 x}. \tag{3.5}$$

This barrier is sometimes called the Schottky–Nordheim (SN) barrier.

Dr. R. Forbes has suggested that, in principle, for metals the FE current density should be given by a so-called "physically complete" F–N -type equation that has the form:

$$J = \left(\lambda_Z a\varphi^{-1}F^2\right) P_F \exp\left[-\nu_F b\varphi^{3/2}/F\right], \tag{3.6}$$

where: ν_F ("nu$_F$") is a correction factor associated with the barrier shape; P_F is a tunneling prefactor; and λ_Z is a correction factor associated with the effective electron supply – this takes into account effects due to integration over states, electronic structure, and temperature. Obviously, this equation contains three correction factors: ν_F, P_F, and λ_Z. The precise forms of these correction factors depend on the physical features of the emission situation under analysis. For example, they might depend on the geometry of the emitter and/or on the chemical nature of its surface.

To relate a measured current–voltage (I_{FE}–V_a) characteristic to a theoretical discussion in terms of local current–density J and local barrier field F, it is necessary to introduce two auxiliary equations. First, a reference point "0" is chosen on the emitter surface. Often this is the point at which the current density is a maximum. Denote the barrier field at this reference point by F_0 and the current density at this reference point by J_0. This field may be related to the applied voltage V_a by the equation:

$$F_0 = \beta_0 V_a, \tag{3.7}$$

where β is the voltage-to-local-electric-field conversion factor and β_0 is the value of β at the reference point. Values for β_0 have to be obtained by electrostatic analysis of the emitter and system geometry.

With real emitters, the local emission current density J varies with position across the surface, and the total emission current I_{FE} has to be determined (at least in principle) by integration. The result of this integration is written in the form:

$$I_{FE} = \int J dA = A_n J_0, \tag{3.8}$$

where A_n is called the "notional emission area." In reality this parameter A_n may be a function of the emitter geometry, the applied voltage, temperature, and possibly other factors; however, it is often approximated as being constant.

Two limiting forms for the conversion factor β are of interest here. In the context of a parallel-plate arrangement β is given by $1/d$, where d is the distance between the two plates. We argue that this expression for the conversion factor is an adequate approximation when the radius of curvature of the tip, r_{tip} is larger than d. In this case, the emission area is treated as a "flat plate," equivalent to the configuration in a parallel-plate capacitor.

For a field emitter with apex radius r_{tip} it is conventional to write β_0 in the form:

$$\beta = \frac{1}{k_f \cdot r_{tip}}, \tag{3.9}$$

where k_f is referred to as the "shape" or "field reduction" factor. When the field emitter is distant from anode, *i.e.* $d >> l$ (l being the length of the conducting shank of the emitter), it is assumed that k_f is constant for small variations in d and that k_f ranges between 5 and 8 (as determined by Sakurai & Müller, 1973, 1977). These values originate from the exact result for a tip with hyperboloidal geometry introduced by Erying, Mackeown, and Millikan (1928) (more recently confirmed by Zuber, Jensen, & Sullivan, 2002), who found $k_f = (1/2)\ln(4d/r_{tip})$. This assumes that the planar anode is located at the focal plane of the hyperboloid and neglects the contribution from the shank. It is one of the major tasks of NFESEM to find a real, accurate value for the field reduction factor and to relate it to electrostatic models.

3.2 Vertical Resolution

If the parallel-plate model is valid, the electric field is inversely proportional to the separation gap; F is proportional to $1/d$, assuming the applied voltage

$V_a{}^2$ is held fixed. Accordingly, Eq. (3.6) with Eq. (3.8) becomes:

$$I_{FE} = A_n \lambda_Z a\varphi^{-1} \left(V_a/d\right)^2 P_F \exp\left[-\nu_F b\varphi^{3/2} d/V_a\right]. \tag{3.10}$$

Therefore a reasonable estimate of the I_{FE} variations can be predicted using the so-called "F–N coordinates:"

$$\ln\left\{I_{FE}/V_a^2\right\} = \ln\left\{A_n \lambda_Z a\varphi^{-1} P_F/d^2\right\} - \nu_F b\varphi^{3/2} d/V_a. \tag{3.11}$$

It follows that the variations in FE current are approximately:

$$\delta \ln\{I_{FE}\} = \frac{\delta I_{FE}}{I_{FE}} \approx -\nu_F b\varphi^{3/2} \left(\frac{d}{V_a}\right)_F \cdot \frac{\delta d}{d}. \tag{3.12}$$

In a first-order approximation, we can set the Schottky–Nordheim correction factor to unity $\nu_F \approx 1$ (Forbes, 2006: note that this is only in the exponential term of I_{FE}). Furthermore, $b\varphi^{3/2}$ is approximately 65 V/nm; therefore:

$$\frac{\delta I_{FE}}{I_{FE}} \approx 65\ \text{V/nm} \cdot \left(\frac{d}{V_a}\right)_F \cdot \frac{\delta d}{d}. \tag{3.13}$$

A typical tip-sample separation gap for a high resolution NFESEM image is $d = 25$ nm, which we have achieved on a W (110) substrate (see Kirk, 2010a) with $V_a \leq 60$ V. This surface exhibits a single atomic step height of 0.2 nm, as we have confirmed by subsequent STM measurements. If we assume similar parameters, then $\delta d/d \sim 0.2/25 = 0.8\%$. Therefore, it follows that we can expect FE current variations $\delta I_{FE}/I_{FE}$ of around 20% (for $V_a = 60$ V). With modern electronics, it is easy to measure this kind of variation when I_{FE} is in the nanoampere range.

It is also important to note that increasing the local electric field reduces the FE current variations. Moreover, if we examine Eq. (3.10), it is apparent that the FE current varies exponentially with the distance, $I_{FE} \sim \exp(-ad)$, which is akin to the variation of tunnel current in STM. The ideal NFESEM measurement consists of achieving a high FE current by applying a low local electric field (or voltage), which is also in agreement with the minimization of the SE deflection by the strong electric field between the tip and sample.

[2] The applied voltage, V_a, is a difference between Fermi levels, so it does not vary along the surface. Accordingly, V_a is the difference in classical electrostatic potential. However, this can vary along the surface, if the local work function varies.

In the above argument we have used the parallel-plate approximation. The FE current is more sensitive at small d when we use the hyperboloidal approximation, which is a better approximation when $r_{tip} \rightarrow d$. Of course, this assumes that the FE current is stable enough to measure these variations in both cases.

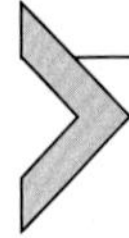

4. PRIMARY ELECTRON BEAM GENERATION

4.1 Emitter Preparation

In a technique like NFESEM, the ability to produce "sharp" and symmetric field emitters is essential; in fact, a good tip fabrication can mean the difference between a successful and a failed measurement. Our field emitters are fabricated from cylindrical 99.98% (metals basis) polycrystalline tungsten (W) wires – from Alfa Aesar – with a nominal diameter of 0.250 mm. The choice of this material is motivated by the correlation between price, quality, and "sharpness" of the etched W-wire. Obviously other W-wire types can be successfully employed, as presented earlier in Kirk, Scholder, De Pietro, Ramsperger, and Pescia (2009), which reports experimental results of single crystal tungsten manufactured tips of (100)-, (111)-, and (310)-orientations used for NFESEM.

Our current procedure is mainly divided into two parts: a manual preparation in laboratory conditions; where, the W-tip is effectively produced via an electrochemical etching technique, also called *ex-situ* preparation, and a heat treatment executed inside a UHV system and therefore called *in-situ* preparation. After these two steps the field emitter is ready for both NFESEM and STM.

Ex Situ Preparation

The first stage of tip preparation begins by "smoothing" the outer layer of a 0.25 W-wire, as the surface of the wire tends to be rough and covered by minor contaminants. This is executed by attaching approximately 5 cm of W-wire to a drill motor and polishing the surface using, preferably, two or more different polishing papers of various roughnesses. Typically, we start with the rough paper and end with the finer one. After about 5 minutes, the wire end begins to fork and is ready to be used. Next, the desired length is manually cut and spot-welded to the tip holder. This tip holder is mainly made from titanium; however a MACOR tube is sometimes used to reduce the electric field generated at the scanner head (see Kirk, 2010a).

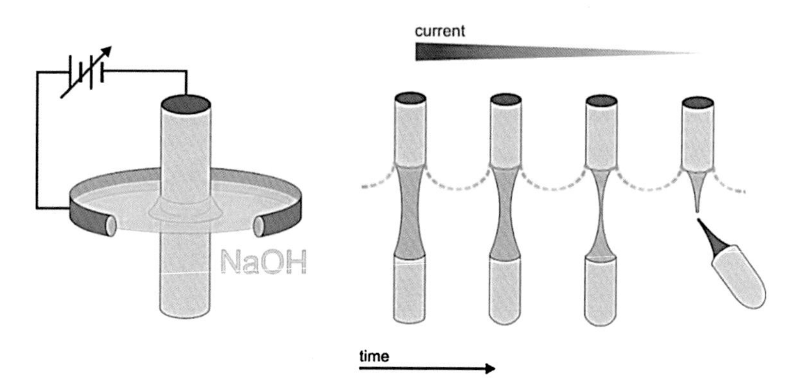

Figure 6 Schematic representation of the etching process. The left side shows a sketch of the initial setup of the etching station. The right side demonstrates the etching behavior with respect to the time. Reprinted from Advances in Imaging and Electron Physics, Vol. 170, Zanin, D. A., Cabrera, H., De Pietro, L. G., Pikulski, M., Goldman, M., Ramsperger, U., Pescia, D., & Xanthakis, J. P., Chapter 5 – Fundamental Aspects of Near-Field Emission Scanning Electron Microscopy, 227–258, Copyright 2012, with permission from Elsevier [OR APPLICABLE SOCIETY COPYRIGHT OWNER].

Afterwards the tip holder is mounted on the support of the tip-etching station, in order to start forming the W-tip. The sharpening of the tip is done through electrochemical etching; where, the W-wire acts as an anode, a platinum ring surrounding the wire serves as a counter-electrode (see left side of Fig. 6), and a 5 mol/L NaOH solution is used as an electrolyte. It is important to note that the effective etching occurs in the region near the air/electrolyte interface. The fact that a small concave meniscus will form around the tip, when it is submerged, will significantly increase the efficiency of the etching. Furthermore, because the reaction is lower at the exact interface between the air and the electrolyte, the meniscus around the W-tip will also contribute to design the shape curvature of the W-tip. During the etching process, the effective radius of the anode (the W-wire) reduces (see right side of Fig. 6), subsequently increasing the resistance of the circuit. The current flowing through the circuit is thus expected to decrease constantly up to the point where it "drops-off" to zero. This point, situated normally between 1.5 mA and 2.0 mA, coincides with the point where the cut off effectively happens, meaning that the tip has successfully etched. In order to avoid further etching, the circuit is "switched off" with the help of an automatic switch-off control. The tip "sharpness" depends on the speed of the switch, so faster switches produce sharper tips.

In the pre-etching phase the W-wire is precisely aligned to be perpendicular to the platinum ring surface and pointing towards its center using

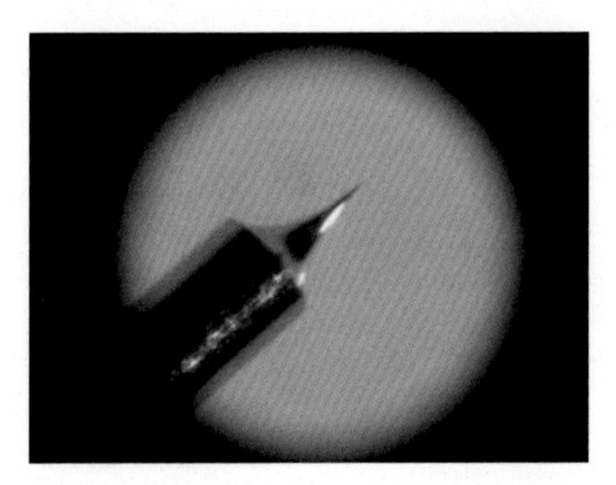

Figure 7 Optical micrograph of the *ex situ* prepared field emitter. The etched portion has a length of about 1–2 mm. Reprinted from Advances in Imaging and Electron Physics, Vol. 170, Zanin, D. A., Cabrera, H., De Pietro, L. G., Pikulski, M., Goldman, M., Ramsperger, U., Pescia, D., & Xanthakis, J. P., Chapter 5 – Fundamental Aspects of Near-Field Emission Scanning Electron Microscopy, 227–258, Copyright 2012, with permission from Elsevier [OR APPLICABLE SOCIETY COPYRIGHT OWNER].

micrometer screw positioning. After this calibration of the etching stage, which determines the symmetry of the tip, the wire is submerged approximately 2 mm into the electrolyte (depicted in the left side of Fig. 6) and, under the action of a cell voltage of 4.38–4.50 V, the pre-etching reaction begins. After this phase, the tip is blunt but symmetric. The need of the pre-etching is to define the starting position ("zero point" for the W-wire) for the next phase.

The "effective" etching begins with the further submersion of the tip over 1 mm and the process is repeated. At the first sight, the current flowing through the circuit behaves in the same manner as during the pre-etching phase; nevertheless, if the etching of the tip is working properly, a "smooth" and constant decrease in the current can be observed. After the "cut-off", meaning that the process is finished, the tip has to be removed together with the tip-holder and immediately washed in deionized water, which has to be pre-heated to $\sim 60°C$, for about one minute. A further cleaning with acetone may be performed to remove the residual drops of water and allow for quick-drying. The manufactured W-tip can now be examined under an optical microscope. If everything was done properly the tip should look sharp and symmetric; however the end of tip should not be visible, as the apex radius is beyond the limit of an optical microscope (see Fig. 7).

In Situ Preparation

The *In Situ* preparation is directly following the *ex situ* preparation and is a necessary procedure to remove the additional surface contaminants on the

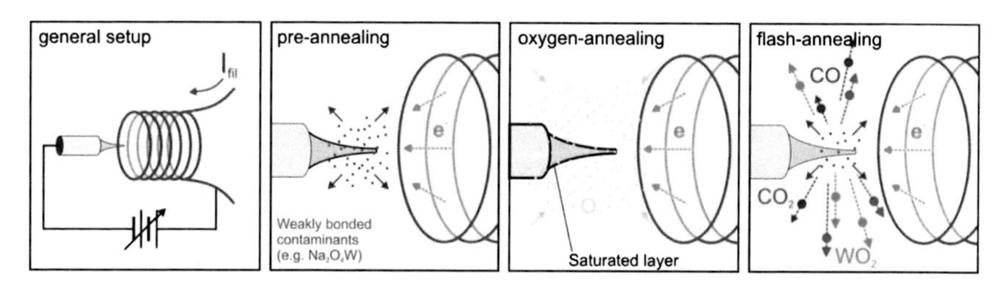

Figure 8 Schematic representation of the annealing process. Reprinted from Advances in Imaging and Electron Physics, Vol. 170, Zanin, D. A., Cabrera, H., De Pietro, L. G., Pikulski, M., Goldman, M., Ramsperger, U., Pescia, D., & Xanthakis, J. P., Chapter 5 – Fundamental Aspects of Near-Field Emission Scanning Electron Microscopy, 227–258, Copyright 2012, with permission from Elsevier [OR APPLICABLE SOCIETY COPYRIGHT OWNER].

finished W-tips. On the one hand, most of the contamination is related to the fact that the *ex situ* procedure is executed in ambient conditions, allowing for the oxygenation of the tungsten surface, *i.e.* tungsten trioxide WO_3. On the other hand, the polycrystalline tungsten wire used is not pure; it contains some tungsten carbide WC. The aim of the procedure explained below is to remove these impurities from the field emitter. The method used for this preparation is divided in three parts (see Fig. 8):

- pre-annealing, where the field emitter is pre-heated and the weakly bound impurities are removed;
- oxygen-annealing, which is used to remove carbon;
- flash-annealing, where the field emitter is heated to remove all contaminants and oxides on the field emitter.

Note that no field is applied to the emitter during these preparation stages.

Before annealing, the first step is to move the manufactured field emitter inside the UHV system, as soon as possible. This is done by introducing the field emitter into the "load-lock," which is the fore-chamber of UHV systems. Once a suitable vacuum condition is reached, the field emitter can be transported in the preparation chamber. Next, the field emitter is put in the center of a 0.125 mm thick, coiled W-filament and oriented along the coil axis (see Fig. 8 leftmost image).

In the pre-annealing part, a voltage of 500 V together with a current flowing through the filament are applied. The current in the filament is slowly, but continuously increased, while taking into account that the pressure of the chamber does not go beyond a pressure threshold of $3 \cdot 10^{-8}$ mbar. This procedure will heat the filament producing a thermal emission of electrons from the filament to the W-tip. This electron bombardment generates a measurable thermal electron emission current; which is increased until the emission current reaches a value of 3.5 mA, and the tip is annealed for at least 45 minutes. During this waiting period the pressure decreases smoothly, indicating that annealing is occurring and the weakly bounded contaminants on the surface are being removed.

When the pressure returns to the mid-to-low 10^{-9} mbar regime, the field emitter is ready for the oxygen-annealing, which is performed by introducing oxygen inside the chamber such that the pressure in the preparation chamber increases to a constant value of $4 \cdot 10^{-8}$ mbar. Under these conditions the field emitter is annealed for an additional 20 minutes. The residual carbon contamination on the surface will then react resulting in surface CO and CO_2; during this procedure, the very first thin layer is completely saturated from the oxygen, and the emission current is kept at 3.5 mA.

Afterwards, once the oxygen-annealing is finished and the pressure stabilizes, the field emitter is flashed for one minute at an emission current of 5.5–6.0 mA depending on the quality of the vacuum inside the chamber. During this treatment, the field emitter reaches a temperature of about 2000°C, at which point the first saturated tungsten layer is removed in accordance with the following reaction:

$$2WO_3(s) + W(s) \rightarrow 3WO_2(g). \tag{4.1}$$

The tungsten-dioxide sublimates already at 800°C and most of the other contaminants will sublimate around 1100°C. It is very important that the W-tip is not melted, even partially, during the flashing phase. This causes a "bulbing" effect at the emitter apex, which will vary the FE characteristics.

Following the preparation of the tip we perform a FE test, aimed at checking the tip sharpness (later, in the measurement chamber, a more careful test can be conducted, if necessary). This preliminary procedure is performed as follows:

1. The tip is brought near (< 5 mm) a conducting sphere (see Fig. 9A) that is held at ground potential (located just above the filament used for electron bombardment in Fig. 8;
2. A negative potential is then applied to the tip inducing a positive bias on the conducting sphere; hence the field emitted electrons from the tip are collected via the conducting sphere;
3. The tip voltage is increased until the FE current reaches exactly 0.1 nA, adjusting the distance accordingly;
4. Step 3) is repeated for various FE currents at the same distance used in the previous step, generating a $I_{FE}-V_a$ curve.

4.1.1 F–N Plot Analysis of Experimental Data

In order to interpret the I_{FE}–V_a data, it is customary to convert it into so-called "F–N" coordinates. This is performed by expressing Eq. (3.10) as:

$$\ln\left\{I_{FE}/V_a^2\right\} = \ln\left\{A_n \lambda_Z a \varphi^{-1} \beta^2 P_F\right\} - \nu_F b \varphi^{3/2}/\beta V_a. \tag{4.2}$$

The plot of $\ln\{I/V_a^2\}$ against $1/V_a$ is known as an "F–N" plot and should yield a (nearly) straight line, assuming a planar model (*i.e.* conventional F–N theory), if the voltage dependence of all correction factors and A_n are disregarded.

Experimental current–voltage FE data is usually analyzed in the form of an F–N plot with a straight line fitted through the data points. The equation of the fitted line is generally given by:

$$Y_V \equiv \ln\left\{I_{FE}/V_a^2\right\} = \ln\{R_V\} + S_V/V_a; \tag{4.3}$$

where S_V is the slope of the fitted line and $\ln\{R_V\}$ its Y_V-intercept. Comparison of this equation with the theoretically predicted curve Eq. (4.2) gives expressions for S_V and R_V:

$$S_V = -\sigma b \varphi^{3/2}/\beta; \quad R_V = \rho A_n a \varphi^{-1} \beta^2. \tag{4.4}$$

The generalized slope and intercept correction factors, σ and ρ, are functions that vary with the field and hence the applied voltage.

The general definition of σ is $\sigma = (\varphi^{3/2}/\beta)^{-1} \partial \ln\{I/V_a^2\}/\partial(1/V_a)$. The values of σ are tabulated for the Schottky–Nordheim barrier, but no exact values are known for a specified setup using the physically complete F–N equation. For our purposes the value $\sigma = 0.95$ should be a useful approximation (Forbes & Deane, 2010). The work function, φ, of the tungsten tips depends on the crystal facet from which the electrons escape. Most of our tips are made from polycrystalline tungsten and usually have the (110) crystal plane at the apex. This plane has the highest work function with 5.25 eV. However, most electrons are emitted from adjacent the (111) plane, which has a work function of 4.47 eV. The effective work function for our tips can be estimated to be $\varphi = 4.5$ eV (Melmed, 1991). The approximation of k_f as 5 introduces an error of roughly 25% in this factor. The errors in φ and σ are so small compared to this value that they can be neglected for our calculations. With the slope, S_V, obtained from the field emission data and all constants known, Eqs. (3.9) and (4.4) allow for the calculation of the effective emission radius of the tips (see Table 4 (Appendix B) for typical values):

$$r_{\text{tip}} = S_{\text{V}} \left(-\sigma b \varphi^{2/3} k_{\text{f}}\right)^{-1}. \tag{4.5}$$

The intercept of the fit with the Y_{V}-axis, on the other hand, gives access to the work function of the emitter material. When ρ and r_{tip} are known, φ can be obtained using expression Eq. (4.4) for R_{V}. In most applications ρ is set to unity, which is how we apply this formulation.

4.1.2 Surface Contamination and FE Current Stability

The $I_{\text{FE}}-V_{\text{a}}$ curve data allow us to determine the effective emission radius by implementing a linear (Kirk, Ramsperger et al., 2009) or a quadratic (Kirk, Scholder et al., 2009) fit. However, it is important that we examine the surface contamination problem of the FE emitters. Initially, our tip preparation did not include oxygen-annealing, as we believed that the tip was simply too small for the omnipresent carbon to cause any significant problems. In fact, many perform STM using W-tips without oxygen-annealing or any heat treatment at all. However, the NFESEM strongly depends on the quality of the field emitter; therefore, it is pertinent that the electron source is properly characterized. We have shown that the vertical resolution of the NFESEM depends on the ability to measure small currents in the nanoampere range (see Section 3.2). The signal-to-noise ratio of the current must be higher than the minimal resolution capabilities, which requires the current – and accordingly the tip – to be stable during the imaging process. A stability test was performed in order to examine the changes in the FE current. This test was done by measuring FE between the tip and the metallic sphere. Fig. 9 shows one of the results of this test, which was measured for more than one hour. The results reveal that after an initial decrease in FE current, the current begins to oscillate by approximately ± 15% around the set current, 2 nA. This is to be compared with commercial cold FE tips, which can be as stable as 2%/hr[3] (Goldstein et al., 2003). The main cause of current instabilities in SEM are due to mechanical vibrations on the tip as well as adsorbates that vary the work function of the emission area. Although our FE tips are stable enough to take an NFESEM image – typical acquisition time is approximately ten

[3] This can only be achieved via an electrical feedback circuit, *i.e.* constant current mode, and UHV conditions $\sim 10^{-9}$ mbar.

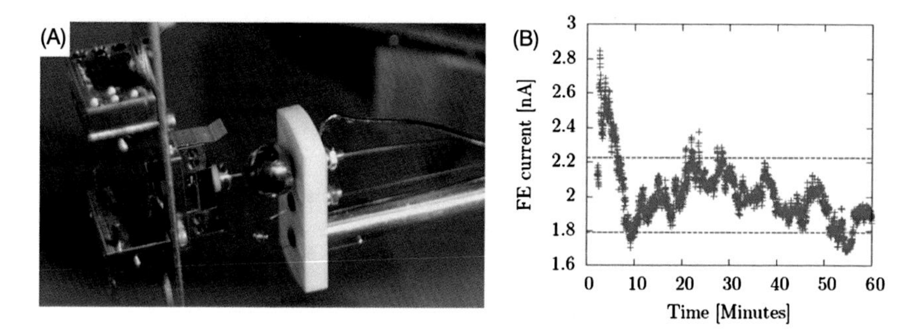

Figure 9 The tip preparation stage. (A) The tip is heated via electron bombardment from the coiled filament (Fig. 8), and subsequently the FE current is measured between the tip and the grounding sphere. (B) The FE stability test show that there are modulations of the FE current under these specific conditions (Kirk, 2010a and Scholder, 2009).

minutes – with atomic vertical resolution (see Section 3.2), it is not as stable as is a high performance W-tip (Tondare, van Druten, Hagen, & Kruit, 2003) in a vacuum more three orders of magnitude worse than our FE test setup, 2×10^{-10} mbar. This led us to believe that the surface of our field emitters may have some contaminants.

We have performed SEM and transmission electron microscopy (TEM), in order to obtain the physical radius of curvature of the tips and their specific shape. Both techniques have been implemented *ex situ*, which allows us to observe the field emitter only upon exposure to ambient conditions (*i.e.* contamination). An optical microscope, *e.g.* shown in Fig. 7, is not an option for imaging because it has a resolution limit of roughly a few hundred nanometers; whereas our tips have radii of the order of nanometers. SEM is a quick and well-known imaging method that does not require flat samples or thin layers. By using TEM, it is possible to evaluate the field emitters from a microscopic point of view; however this requires extra preparation of the field emitter, subsequently adding more contamination.

SEM Imaging

The fully prepared and characterized tips are taken out of the vacuum system just before imaging, to keep oxidation and contamination at a minimum. The tips are mounted on holders developed to be compatible with the SEM and introduced into the vacuum environment of the SEM after only a few minutes in the air. The vacuum conditions in SEM are between 10^{-6} and 10^{-7} mbar during operation. A Zeiss LEO 1530 SEM with a Gemini column is used to perform the measurements. The microscope is

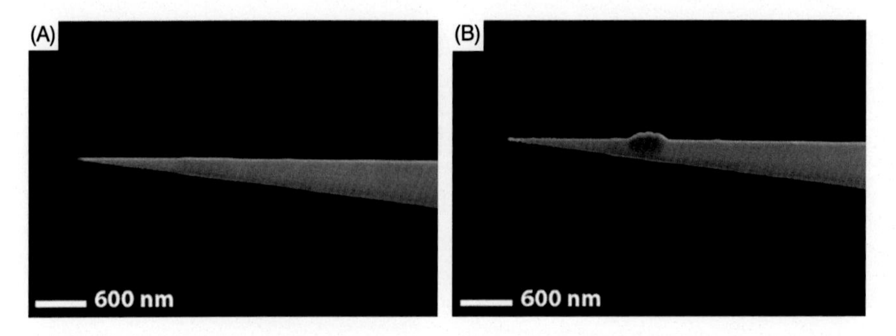

Figure 10 Example of the build-up on the tip. (A) Initial image taken of the field emitter, and (B) subsequent image taken after focusing on the thickened area. The scale is similar in both images. The protruding growth has a width of about 400 nm and a height of a little less than 100 nm. Note that the end of the tip has also deteriorated (Stockklauser, 2010). *Reprinted with permission from Anna Stockklauser.*

equipped with two detectors, one SE in-lens and one external detector (SE2). The in-lens detector only detects low energy SEs, while the detector SE2 also collects some BSEs. Because the SEs are ejected towards the optical axis, the in-lens detector is more sensitive to the surface features. The beam diameter of the instrument is approximately 1 nm. Taking other factors into account this yields a beam energy dependent resolution of about 2–5 nm for a beam energy of 1 kV and 1–2 nm at 20 kV.

Fig. 10A shows an initial image of the tip. One particular spot on the tip was selected in order to focus the electron beam. This location was subject to intensive electron bombardment for several moments during focusing. We have noticed the growth of a lump in the focusing area, which is shown in Fig. 10B. The lump is roughly 450 nm wide and raises slightly less than 100 nm. These dimensions are enormous compared to the tip's apex radius that can also be roughly estimated from the figure. The tip orientation has not been changed during the imaging and the scales in both images are the same. Of course, the build-up has a particularly strong effect on the apex region of the tip. Here, the growing of initially small protrusions was observed. The apex as a whole also grows significantly.

To gain further information about the lump material, we reintroduced the tip into the UHV preparation chamber after SEM imaging and repeated the annealing process. The tip was initially annealed without oxygen (Fig. 11A) and this second annealing procedure (Fig. 11B) was also performed without oxygen. The hope was that the grown material would come off during annealing. As expected, the observed field emission from

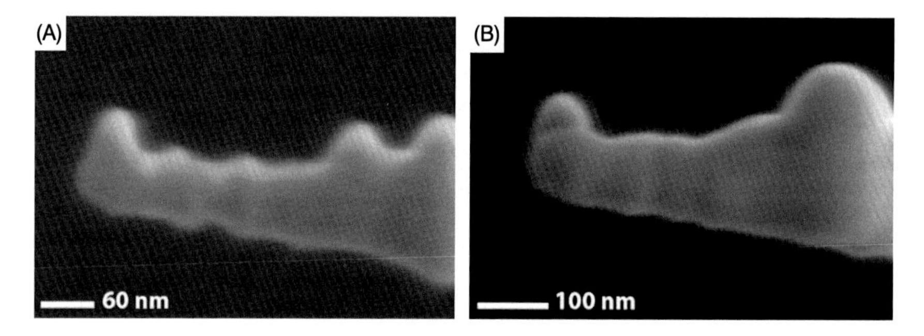

Figure 11 SEM micrograph of the tip apex of Fig. 10. The former image (A) was taken after the electron beam-induced build-up, and the latter image (B) was made after a subsequent high temperature annealing was performed, in order to try to remove the contaminants (Stockklauser, 2010). *Reprinted with permission from Anna Stockklauser.*

the tip after the second annealing suggested a considerably larger emitter radius than that observed after the first.

We have experienced the same problem in the imaging of almost all tips that did not receive an oxygen-annealing. In each case, it was apparent that the electron bombardment was the origin of the growth. A slight change of the sample surface can often be noticed during SEM imaging; however, in the present form the occurrence is very unusual. Therefore, we have attributed this growth to our preparation of the field emitters; in particular, the heat treatment. Apparently, the surface of the electron sources was not tungsten, as we have previously thought, but rather carbonated tungsten. Furthermore, the high energy SEM beam electrons contribute to the surface migration of the bulk carbon, which is still within the tungsten after annealing. This combined with the well-documented SEM beam contamination, caused by breaking hydrocarbon bonds in poor vacuum, appears to grow on this seed layer. The fact that vacuum annealing is not sufficient to remove the lumps is in agreement with this theory.

Assuming that the origin of the observed build-up on the tips is carbon, we have introduced the aforementioned oxygen annealing step in the tip preparation procedure. This was intended to remove the carbon during the preparation, so that it could not come to the surface during imaging. For the tips that were annealed in an oxygen atmosphere the tungsten texture was much more visible. The polycrystalline grains of three different tips are shown in Fig. 12. A possible explanation is that without oxygen annealing, a layer of the buildup already forms during the annealing process, as this heating method is also based on electron bombardment. The oxygen

Figure 12 SEM micrograph of three different tips that have been annealed in an oxygen atmosphere. The polycrystalline grains are clearly visible in each (Stockklauser, 2010). *Reprinted with permission from Anna Stockklauser.*

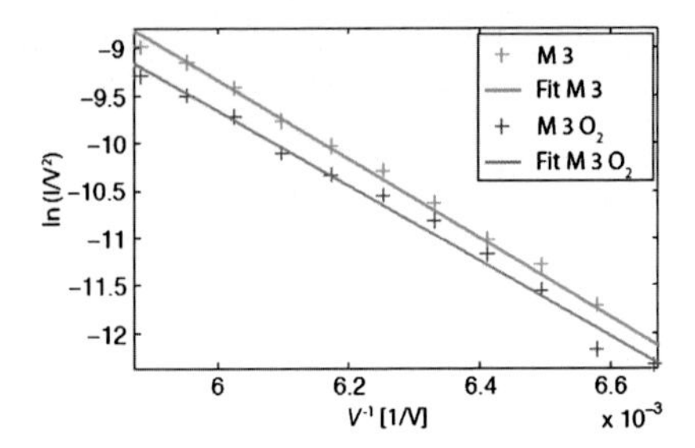

Figure 13 F–N plots of tip M 3 before and after oxygen annealing. As expected the Y_V-intercept of the fit for M 3 O_2 is lower and suggests a higher work function than that for M 3 (Stockklauser, 2010). *Reprinted with permission from Anna Stockklauser.*

annealing could prevent the formation of this layer so that the grain of the tips would lie free when the tips are introduced in the SEM.

Annealing the tips in an oxygen atmosphere was introduced to remove the omnipresent carbon from the tip surface, changing the emitting material from tungsten carbide to tungsten. Thus an effect of the annealing method on the work function is expected. Eq. (4.4) states how the Y_V-intercept of an F–N plot is related to the work function. At 3.38 eV (Vida, Josepovits, Gyor, & Deak, 2003), WC has a lower work function than pure W-tips; therefore, it is expected that the F–N curve has a lower Y_V-intercept, after oxygen-annealing, *cf.* oxygen-free heat treatment. The exponent in the F–N equation also has a factor from the work function, but the slope of the fit should roughly remain the same. We have first annealed a tip (called M 3) without oxygen. Then the tip was cooled and we measured a $I_{FE}-V_a$ curve, see Fig. 13. The tip was subsequently annealed with oxygen and data for another F–N plot was recorded.

It is not expected that our fit can yield the actual work function, as the measurement is not accurate enough and all constants were considered the same for both materials. In the error calculations, only the errors in the two Y_V-intercept values are taken into account. Nevertheless, the example shows how oxygen-annealing increases the calculated work function because it changes the emitting material from WC to W.

TEM Imaging

Controlling and possibly changing our tip preparation techniques required more knowledge of the contamination and how we could effectively correct for it. An interesting preliminary way to check the material property of the outer layer is provided by high resolution TEM (HRTEM), which allows us to study the shape and chemical makeup on an atomic scale.

All of our field emitter characterization techniques allowed us to determine the most suitable manner in which the field emitter should be prepared. We have learned many structural and chemical aspects of electron sources that are often neglected in SPM. Eventually this has enabled us to minimize the amount of surface carbonation of the polycrystalline W-tips, while reducing the sequential tungsten-oxide layer. It is especially important to restrict the amount of surface oxides as this can completely inhibit FE, which we have also observed in our analysis. In order to fully remove the contaminants, the tip is optimally heated higher by turning up the bombardment current to 5.8 mA for 1 minute, while keeping the pressure below a maximum threshold of approximately $7 \cdot 10^{-9}$ mbar. Additional heating, *i.e.* flashing, has resulted in the melting of the field emitter, as confirmed via subsequent SEM and TEM imaging. One of the tips that were optimally prepared is shown in Fig. 14. The edge of the tungsten portion of the tip is outlined in white; note that this trace has an ellipsoidal shape, and it is more accurate near the apex of the emitter. A tungsten-oxide layer – less than one nanometer thick – can be observed immediately after the ellipse. This is followed by a large amount of carbon that surrounds the entire field emitter, but it is concentrated at the apex. The spatially resolved spectra in Fig. 15 show the chemical composition of the apex. Fig. 15B reveals that the carbon signal is, by far, the most dominant. Furthermore, the transition from a dominant oxygen signal to a more pronounced tungsten signal indicates that the oxide layer is still present; however, it is confined to less than a nanometer (in agreement with the HRTEM micrograph in Fig. 14).

Our high resolution imaging of the electron sources was not only to assess the structural and chemical nature of the emitter, but to determine the physical (or actual) radius (r_{phys}) at the apex as well. This is somewhat subjective, as the radius is determined by an arbitrary circle inscribed at

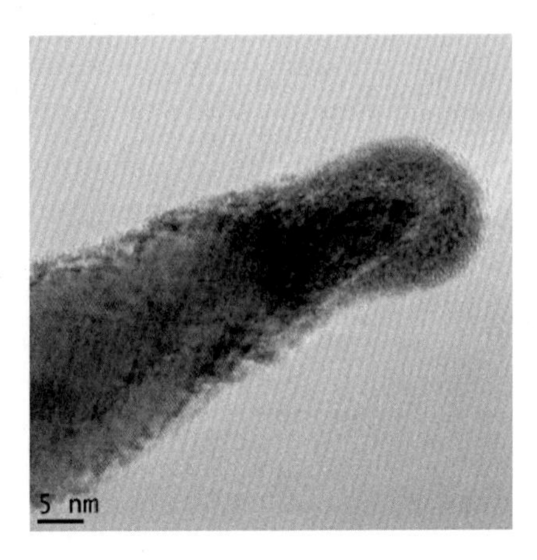

Figure 14 TEM micrograph of a "sharp" field emitter with a thin layer of tungsten oxide (light area just outside the ellipse) and electron beam-induced carbon contamination (darker region after the oxide layer). The white ellipsoidal outline shows the extent of the purely tungsten part of the tip that is more accurately marked near the apex. Reprinted from Advances in Imaging and Electron Physics, Vol. 170, Zanin, D. A., Cabrera, H., De Pietro, L. G., Pikulski, M., Goldman, M., Ramsperger, U., Pescia, D., & Xanthakis, J. P., Chapter 5 – Fundamental Aspects of Near-Field Emission Scanning Electron Microscopy, 227–258, Copyright 2012, with permission from Elsevier [OR APPLICABLE SOCIETY COPYRIGHT OWNER]. *Image provided by Magdalena Parlinska-Wojtan at the Center for Electron Microscopy, EMPA.*

the apex. We use SEM to image the tips due to its characteristically high depth of focus, which allows for the imaging of the emitter shank. The resolution capabilities of TEM enable us to determine the apex shape with high precision, since it maps a 2D slice of the considered sample. The results from these measurements will allow us to make a comparison with a corresponding effective emission radius (r_{eff}) – *i.e.* the tip radius expected from the observed FE – obtained by measurements of $I_{FE}-V_a$ curves at macroscopic distances and application of F–N theory (see Section 4.1.1). We have made assumptions on certain parameters, such as the shape factor k_f describing the apex geometry; however the results of the comparison between the r_{phys} and r_{eff} appear to be comparable, at least for $r_{tip} > 10$ nm. These findings are summarized in Appendix B. Bear in mind that standard linear, *i.e.* planar, F–N plot interpretation is not valid for $r_{tip} < 10$ nm, and an alternative method must be implemented (discussed in the following section). Therefore, field emitters with a radius of curvature, as in Figs. 14 and 15, are expected to exhibit a slightly curved F–N behavior (see Kirk, Scholder et al., 2009).

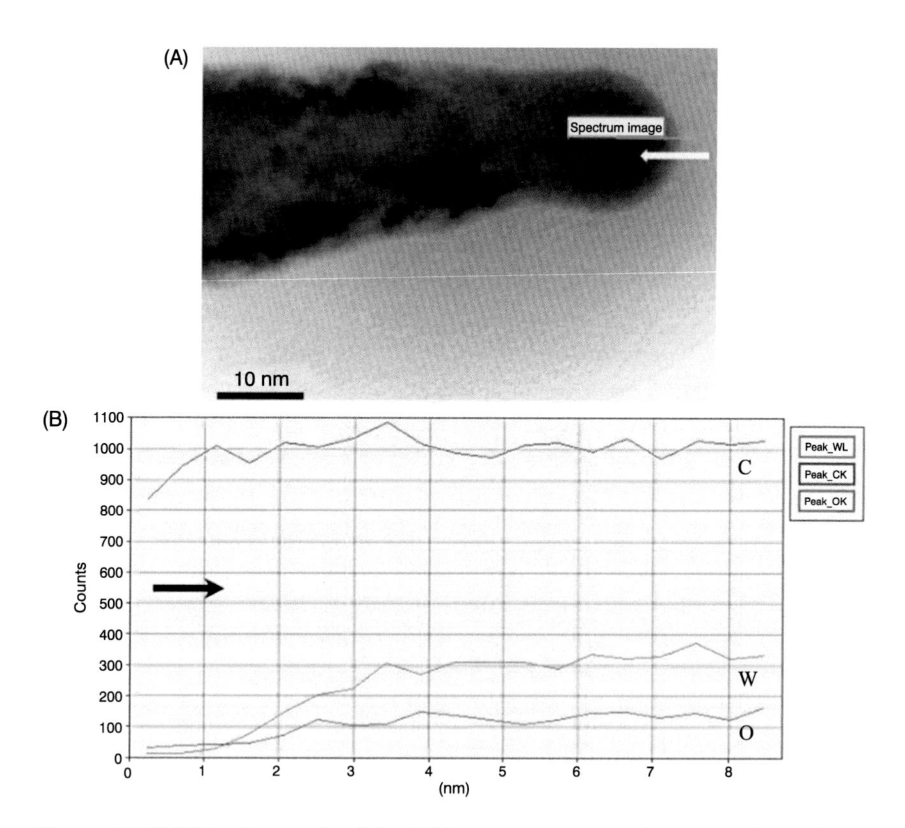

Figure 15 (A) TEM micrograph of the field emitter in Fig. 14. The image above shows where the (B) spectral line scan was made; in order to determine the chemical composition of the apex. The arrows indicate the direction of the spectral scan. Reprinted from Advances in Imaging and Electron Physics, Vol. 170, Zanin, D. A., Cabrera, H., De Pietro, L. G., Pikulski, M., Goldman, M., Ramsperger, U., Pescia, D., & Xanthakis, J. P., Chapter 5 – Fundamental Aspects of Near-Field Emission Scanning Electron Microscopy, 227–258, Copyright 2012, with permission from Elsevier [OR APPLICABLE SOCIETY COPYRIGHT OWNER].

4.2 Field Emission from Non-Planar Surfaces

For the practical application of F–N theory, *i.e.* when the equation is expressed in $I_{FE}-V_a$ coordinates, the geometry of the tip is incorporated via the applied voltage-to-local electric field conversion factor, β. For tips of fairly large emitter radius (> 50 nm) this holds true for the tip geometry: In the parallel plate arrangement that is assumed for these emitters the field is constant everywhere between the plates and voltage-to-local-electric-field conversion writes $F = \beta V_a$ with a constant β value. Prof. P. Cutler et al. (see He et al., 1991; Cutler et al., 1993) have explored the breakdown of standard F–N theory for emitters with a high radius of curvature and have shown that small-radii emitters exhibit deviations from linear F–N curve behavior. They have clearly shown that the standard F–N-type equation does not apply to situations where the emitter is sharply curved. Rather, the correction factor ν_F, in the physically complete F–N-type equation must be replaced by something more general than the mathematical function υ. Modifications to the pre-exponential will also be necessary. It is important to mention that deviations from F–N plot linearity, for sharp emitters, are related to neither space charge effects nor surface roughness. Usually, the current density is lower than the threshold for space charge effects, which is generally around $J > 10^{10}$ A/m^2.

Even with classically flat emitters, there are effects that would cause F–N plots to be slightly curved. However with sharp emitters, the main effect that causes non-linearities in the F–N plot is the fall-off in the strength of the electrostatic field with distance from the emitter surface. For this reason, revisions of the F–N equation are required for nanometer emitters, which must include an asymmetric potential barrier that varies with the tip radius and the polar angle (see Section 4.3 for a detailed discussion).

P. Cutler's method involves numerical calculations for specified emitter geometries, *e.g.* hyperboloid and cone, which influence the barrier shape. The resultant F–N curves were fitted to the following equation:

$$AV_a^2 \exp\left(-B/V_a - C/V_a^2\right); \tag{4.6}$$

where A, B, and C are constants that depend on material and geometrical properties.

G. Fursey has also prescribed a barrier model, based on a spherical emitter (Fursey, 2005). In effect, he expands on the motive energy (see Kirk, 2010a) to include the field fall-off. This leads to the expression:

$$M(x) = h^* - eFx\left(\frac{r_{\mathrm{tip}}}{x + r_{\mathrm{tip}}}\right) - \frac{e^2}{16\pi\varepsilon_0 x}. \tag{4.7}$$

It follows that defining a suitable potential barrier may enable the determination of microscopic information about the emitter from I_{FE}–V_a characteristics (refer to Edgcombe & De Jonge, 2007; Kirk, Scholder et al., 2009). Recently, a FE curved-surface theory (detailed in Edgcombe & De Jonge, 2007) was developed to explore the emission properties of a carbon nanotube (CNT). This demonstrated the theory's ability to deduce the work function of the surface, along with the effective emitter radius, the surface field, the effective solid angle of emission and the supply factor from measured characteristics. Dr. C. Edgcombe has derived a direct relationship between the curvature of an F–N plot and a function describing a hemispherical barrier, as it is varied along the emitter surface:

$$\frac{\partial S/\partial V_{\mathrm{a}}^{-1}}{S\cdot V_{\mathrm{a}}}=x\frac{\partial^2 f_1/\partial x^2}{\partial f_1/\partial x}; \tag{4.8}$$

where S is the rate of change of the exponent of the measured current dependence with V_{a}^{-1}, x is the ratio of the minimum barrier thickness to the emitter radius of curvature, and f_1 is the barrier integral. The curvature of the F–N plot is determined via a quadratic fit of the form:

$$\ln\left(I_{\mathrm{FE}}/V_{\mathrm{a}}^2\right)=AV_{\mathrm{a}}^{-2}+BV_{\mathrm{a}}^{-1}+C; \tag{4.9}$$

where A, B, and C are fitting parameters (different from the ones used above). Note that this is the same dependence observed by Cutler et al.; hence this method is appropriate for emitters that have a radius of curvature equal to (or less than) the barrier width. The fitting parameters evaluated at an arbitrary V_{a}^{-1} are used to estimate a value for the emitter radius using the following formula:

$$\sqrt{\varphi}\cdot r_{\mathrm{tip}}=(c_2ex)^{-1}\frac{-SV_{\mathrm{a}}^{-1}}{\partial f_1/\partial x}, \tag{4.10}$$

here $c_2=\left(\frac{4\sqrt{2m_{\mathrm{e}}}}{3e\hbar}\right)$.

This generated a reasonable estimation of the associated field emitter properties using only the curvature of the F–N plot and the energy distribution of the field-emitted electrons. The work function and subsequently the emitter radius can be determined more precisely using the parameter d_0, which is a parameter (sometimes called the "decay width") that relates to the rate of change of (the natural logarithm of) the barrier penetration coefficient D with forwards energy W, taken at the Fermi level E_{F}. A similar procedure was also applied to emitters used in the present microscope.

Finally, one should mention the very recent work by Xanthakis and Kyritsakis, in which the authors obtained a generalized Fowler–Nordheim type equation for the field emitted current from nanometric emitters (Kyritsakis & Xanthakis, 2015). Correction factors – for the non-linearities in the potential – in both the exponential and pre-exponential terms were analytically deduced and given in the paper. Comparison with the data of three eminent experimental groups gives support to the theory. It must be stressed though that the problem of the lateral resolution of the NFESEM (see below) is a harder problem in that it requires the accurate calculation of the distribution of the current spatially.

4.3 Lensless Focusing

Although the NFESEM has already demonstrated its atomic vertical sensitivity via the correlation between the minimal variations in the FE current, due to the topographically-induced variations in the local electric field, F, the most compelling result has yet to be discussed. In SEM the wavelength of the electrons – determining the diffraction limit for the lateral spatial resolution – is on the order of 0.1 nm or less. Accordingly, the lateral spatial resolution is effectively defined by the actual lateral length scale onto which the electron beam can be focused. Focusing the electron beam, technically, presents the most difficulty for any SEM-type instrument. In SEM with a remote source, this problem is solved by a sophisticated sequence of electrostatic and magnetic electron lenses. In lensless NFESEM, one has to rely on the geometrical and physical properties of the field emitted electron beam. We have already discussed how the NFESEM overcomes the vertical resolution diffraction limitations of approximately 0.4 nm to resolve structures that are 0.2 nm in height, now we will show how the NFESEM can focus[4] the field emitted electrons without any additional components.

Prof. J.J. Sáenz et al. have considered a model consisting of a hyperboloidal emitter cathode and a planar anode in the "near field" emission regime and computed the NFESEM lateral spatial resolution to be (Sáenz & García, 1994):

$$\Delta x \approx 0.7\sqrt{\left(r_{\text{tip}} + d\right) \cdot d}. \tag{4.11}$$

As a result, the lateral resolution dependence on the r_{tip} and d indicate that atomic lateral resolution is not feasible under the conditions investigated by Sáenz. At distances used in the NFESEM experiment, typically d is of the order of a few tens of nanometers, Sáenz estimates a lateral resolution of the order of d. It must be stressed however that Sáenz assumed that the planar anode is located at the focal plane of the hyperboloid and hence his results are limited to values of d that conform to this condition approximately.

[4] "Beam concentrating" is meant by focusing of the field emitted electrons, since the electrons are concentrated along the optical axis, due to the broadening of the barrier for paths located away from the axis.

In the same paper, Sáenz also predicts a possible vertical resolution of the order of less than 0.2 nm, in the NFESEM regime, and demonstrates this performance by experimentally detecting a step edge on a graphite surface (Sáenz & García, 1994).

We have previously investigated how non-planar FE theory (Edgcombe & De Jonge, 2007) can be applied to NFESEM technology, in order to characterize the field emitter (discussed in Kirk, Ramsperger et al., 2009). This strategy has led us to deduce a reasonable r_{eff}; however the resultant effective solid angle is "large," $\sim$ 1.64 sr. Our deduced angle was smaller than the presented solid angle of a carbon nanotube ($\sim$ 2.5 sr) with a similar radius, but it would not result in the lateral resolutions that we have observed (as low as 2–3 nm, Kirk, De Pietro et al., 2009; Kirk, Scholder et al., 2009).

In many models of FE, the field emitter shank is often terminated by a hemispherical cap, *e.g.* Edgcombe's non-planar FE theory. However, our field emitters are more ellipsoidal in shape. Accordingly, we have been working with Prof. J. Xanthakis et al. at the National Technical University of Athens (NTUA) to help us develop a theory that will explain the reason for the discrepancy between the predicted lateral resolution *LR*, of the order of d and the observed resolution Δx – a few nanometers. The NTUA group has formulated a theory that chooses to terminate the field emitter with a hemi-ellipsoidal cap and uses a 3-dimensional JWKB method for the computations (Kyritsakis et al., 2010). They find a self-focusing of the electron beam *LR* that corrects Sáenz's prediction of the *LR* toward the experimentally observed values Δx.

The mechanism of self-focusing is based upon two main ideas. First, the self-focusing is highly sensitive to the sharpness of the tip, which can be quantified as $S = R_1/R_2$; where R_1 and R_2 are the large and small radii, respectively, of the ellipsoid simulating the tip. Second, the actual simulation entails simulating the tip by a stack of these ellipsoids, so as to take proper account of the macroscopic dimensions of the tip shank. This is important in order to determine the actual electric field at the apex. It also allows variations in the tip-to-anode distance d, in contrast to the above mentioned calculation of Sáenz, and finally makes the calculations tractable.

The results for the calculated potential as a function of the angle of deviation from the optical axis – at a well-defined distance of 0.1 nm away from the field emitter surface – are shown in Fig. 16 for different S. The most remarkable characteristic is a sharpening of the potential around the tip axis with increasing tip sharpness. This observation hints to the second concept

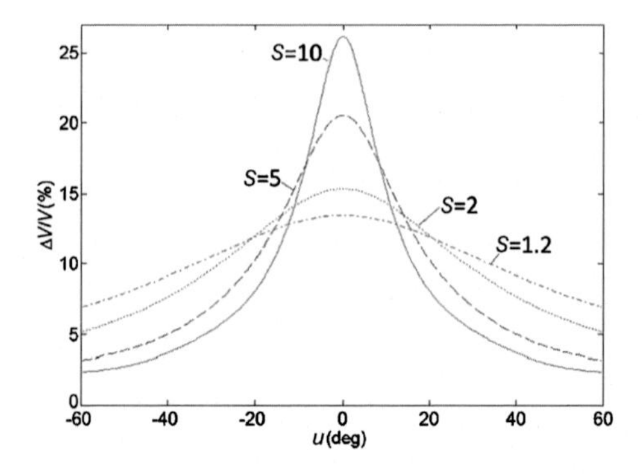

Figure 16 Variation of the electrostatic potential with spheroidal angle (u) at an optical axis distance of 1 Å away from the emitting surface for various values of the sharpness S at a constant $R_2 = 10$ nm (Kyritsakis, Xanthakis, Kirk, & Pescia, 2011). The initial calculations – in Kyritsakis et al. (2010) – were made in the infinite separation regime, *i.e.* $d >> h >> r_{\text{tip}}$; therefore this relationship is valid for most electron gun columns used in SEM, TEM, and other electron beam-based techniques. More recently the simulations have been customized to suit NFESEM parameters, including the image potential configured at $d << h$.

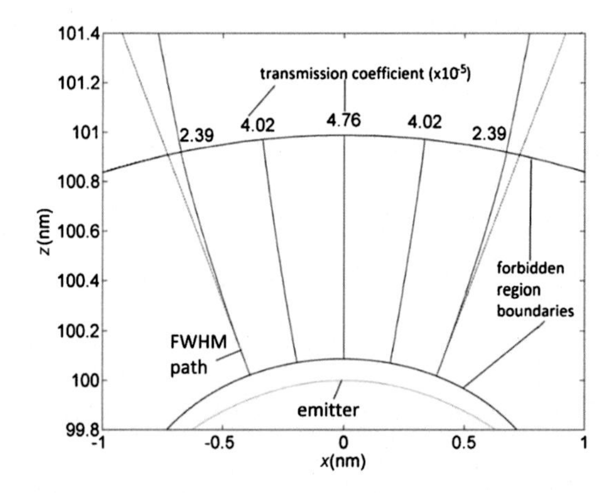

Figure 17 A diagram of the electron trajectories in the barrier potential (Kyritsakis et al., 2011).

behind the self-focusing idea. The field emitted electrons travel through the classically forbidden region, *i.e.* the barrier potential, which also follows the shape of the potential lines given in Fig. 17, on their way to the vacuum. One can imagine trajectories that the electrons travel, which are defined in the classically allowed regions and are described via the Feynman path integral method in the classically forbidden region. In virtue of the highly non-spherical shape of the barrier potential, two important phenomena are observed by numerical computation of the transmission coefficient: i) the effective area of the apex; where a sizeable transmission is expected, is reduced with respect to the standard JWKB one-dimensional model and ii) the transmission coefficients peaks are strongly moved forward. Furthermore, the electron trajectories are being bent in the forbidden zone, see Fig. 17.

The trajectories were also continued to a distance of 10 nm from the surface of the field emitter; where a beam width or *LR* can be defined by a FWHM criterion, *i.e.* the width at which – the *transmission coefficient* drops to half of its value from the direction along the optical axis. If conventional straight electron trajectories – perpendicular to the electrostatic potential – are used, the calculated *LR* is about 8.5 nm. This is in stark contrast to the *LR* for electron trajectories calculated using a 3-D JWKB approximation that show curving of the trajectories towards the optical axis, 4.3 nm. It then follows, as concluded by the NTUA group, that the field emitter sharpness *S* modifies the barrier potential so as to converge the electron trajectories for non-planar surfaces (*i.e.* $r_{tip} \sim L$). The above calculations have been performed for radii R_1 and R_2, which yields a tip curvature of approximately 1 nm. Kyritsakis and Xanthakis later extended the calculations using ellipsoids that fit the actual experimental tips in order to calculate the spot size as a function of the tip anode distance *d* (Kyritsakis & Xanthakis, 2013). Their results showed that at $d = 5$–10 nm a spot size of approximately 2 nm is feasible. The *LR* can be improved by either using a non-planar or "sharp" field emitter, or reducing the local electric field (assuming the same FE current is achieved, which is also in accordance of enhancing the vertical resolution – see Section 3.2).

Bear in mind that this self-focusing effect, based on the implementation of a JWKB approximation is not the only explanation for lensless focusing. Dr. C. Edgcombe proposed solutions to Schrödinger's equation using no approximation for electron emission from a small area in a system of planar electrodes. These solutions are for two sources as a comparison with experimental results for electrons passing through two slits (see Edgcombe, 2010a) and for a single source (see Edgcombe, 2010b). Edgcombe's approach indicate that the field emitted electrons are being concentrated along the axis normal to the cathode surface *cf.* field emitted electrons from a point source ejected into a region of uniform potential. In a final analysis, both J. Xanthakis' and C. Edgcombe's theory must be considered for lensless focusing, as the two do not negate each other.

Note that all of the NFESEM experiments were performed at room temperature; therefore thermal effects on the emitter were not considered. Although it has been determined that the electron beam is only partially coherent at room temperature, cooling the emitter will enhance coherence (Cho, Ichimura, Shimizu, & Oschima, 2004). This can possibly increase NFESEM resolution capabilities.

4.4 Source Characterization

Electrostatic Potential Imaging

As presented in Sections 3 and 4, in order to fully characterize the field emitters it is necessary to understand the proper physics describing field electron emission from a sharp tip. Since the previously mentioned problem with oxygenation of the tip is assumed to be small and thus negligible for FE, the simplest possible way to investigate these properties is to directly observe the electrostatic potential during a field emission test. This was performed by electron holography at the Center for Electron Nanoscopy at the Technical University of Denmark (DTU) together with Dr. T. Kasama in the group of the Prof. R.E. Dunin-Borkowski who is currently the director of Microstructure Research at the Forschungszentrum Juelich and a professor at RWTH Aachen University. The microscope used was a Titan from the FEI company with an electron beam energy of 300 kV in a vacuum pressure of about 10^{-7} mbar. The electron holography set-up and the description of the phase shift reconstruction are described in Midgley and Dunin-Borkowski (2009).

Unfortunately, during the reserved time at the TEM, we continued to have a problem with surface oxides. The main problem encountered was the inability to reach the FE regime using the field emitters investigated, due to an excess of contamination (more precisely oxygen) in combination with the incapability of applying a high voltage between the cathode and the counter-electrode (or anode). Accordingly, the results in Fig. 18B and C must be considered preliminary. In this experiment, the field emitter was placed in close proximity to a planar anode, to which a positive bias was applied. Fig. 18B and C shows the reconstructed phase image of the electron holograms; where the distance between the equipotential lines (white or black) corresponding to a phase shift of 2π. The change in the phase shift, noted by a change in (black/white) contrast, indicates a change in the electrostatic potential. We were able to view the electrostatic potentials around the tip; however, we were not able to observe a measurable FE current.

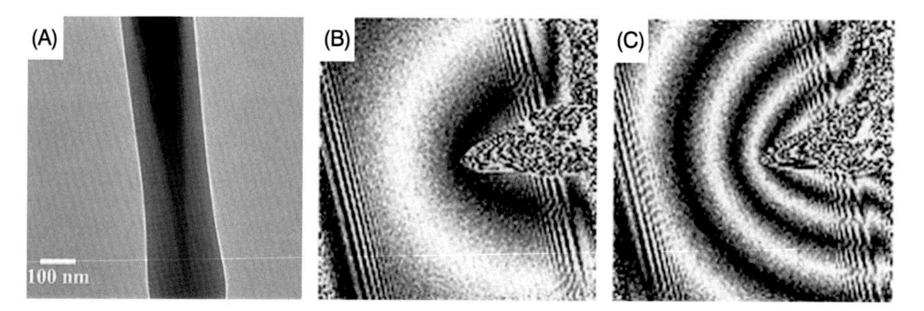

Figure 18 Electrostatic potential reconstruction of the electron holograph. (A) is a TEM micrograph of the field emitter; whereas (B) and (C) are the reconstructed phase image of the electron holograph images recorded by applying a voltage of 0 V and +10 V to the extractor anode, respectively. The dimensions of (B) and (C) are both 500 × 500 nm for the entire frame (T. Kasama and R.E. Dunin-Borkowski, unpublished work).

For this specific tip, we were able to deduce that the contamination was a 300 nm-length × 100 nm-thick layer of oxygen (see Fig. 18A) via EDX. The extra contamination was probably coming from an error during the preparation of the sample, more precisely during the tip annealing. In fact, the observed field emitters were prepared together without considering that, during the oxygen annealing, not only the current field emitter gets oxygenated but also the whole system, gets saturated with oxygen. It is therefore clear that, if a prepared tip stays inside the preparation chamber during the oxygen annealing of another W-tip, a layer of oxygen will be redeposited on the prepared field emitter, rendering it useless for our experiment. Furthermore, the dimensions of the oxygen layer suggest that after the first layer of tungsten-oxide, followed by carbon-oxides, more layers of water have been deposited on the surface permitting this massive growth of the contamination.

Voltage–Distance Characteristics

Three measurable quantities characterize the FE process in the NFESEM regime: the applied voltage V_a, the FE current and the distance d between the apex of the tip and the surface. This section details the electronic characterization of the tip-surface interface, via V_a versus d, and I_{FE} versus d curves. Here we pay attention in particular to measurements of the applied voltage versus distance characteristics, at fixed FE currents (which means at fixed electric fields). The V_a versus d dependence at constant FE current is described by the equation:

$$V_a = F_0\beta^{-1} \quad \text{and} \quad \beta^{-1} = d \rightarrow k_f r_{tip}; \tag{4.12}$$

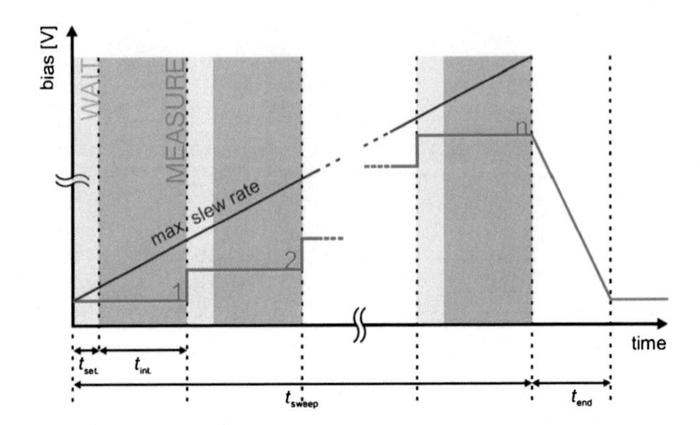

Figure 19 Time characterization of a bias sweep used in the Nanonis Module: **Bias Spectroscopy**. $t_{set.}$ is the settling time; $t_{int.}$ is the integration time over which the point is averaged; t_{sweep} is the total duration of the sweep; and t_{end} is the end settling time, used to return to the starting position after a sweep.

where β is the voltage-to-barrier-field conversion factor. Experimentally, our aim was to use a measured curve in conjunction with the theoretical prediction for the conversion factor, in order to determine physical characteristics of the emitter, such as the radius of curvature of the tip (to be compared *e.g.* with electron microscope images of the apex).

The measurement is done by using the STM with a fixed tunnel current between the tip and sample, *i.e.* CC mode, and slowly increasing the voltage applied between the field emitter and the sample from 0.2 V to 100 V, a so-called bias sweep. The continuous increase of the voltage is translated into a retraction of the W-tip from the sample that is then recorded, defining in this way the V_a–d curve. After some manual tests with the Nanonis software, a semi-automated procedure was performed; in order to improve the quality of the measurements, while increasing the reproducibility of the measurement. This procedure, presented below, is executed with the help of the Nanonis module: **Bias spectroscopy**. This module lets the user perform a bias sweep while measuring more arbitrary channels (*e.g.* position and current) and is typically used to measure I_{FE}–V_a curves. Sweeps are mainly characterized by: the range of voltage to observe; the number of points; the settling time, over which the system can relax after a voltage increase; and the integration time, which represents the period of acquisition of data for each point. Fig. 19 shows the graphical correspondence of the times characterizing bias sweep.

In addition to the voltage range, the number of points and the times described above, Fig. 19 also shows the maximal slew rate that gives the maximal rate at which the bias is allowed to change. This threshold is valid for the whole measurement (*i.e.* before, during and after each bias sweep) and therefore deciding the duration of the end settling time. Additionally it is also possible to define a **Z-control time**, defining a period over which the z-piezo controller is free to move.

It is not possible to know the voltage working range of the investigated field emitter – prior to performing a V_a–d measurement – due to the difficulty in producing similar field emitters. Therefore the measurement is divided in two stages:

1. **test measurement**: one performs a forward sweep that only covers the whole range from 0.5 V to 100 V and, in the best case, needs to be manually stopped, due to a "good" response of the field emitter during the increasing in the voltage;
2. **effective measurement**: one or more forward and backward sweep covering only the allowed voltages.

The bias spectroscopy module produces a graphical output after each sweep and is capable of simultaneously showing a maximum of ten sweeps, which is provided for each file containing all the measured data. These files can be easily processed via standard, technical computing software (*e.g.* Matlab). In order to get an idea of the measurement progress, it is important to also control the time dependence of the distance during the measurement. This can be done by using the **long term chart**, which also provides the possibility to store the information at the end of the measurement. There are references of these characteristic curves in Kirk (2010a, 2010b), as well as Zanin et al. (2012).

In addition, the NTUA group performed simulations of the beam width, here referred to as lateral resolution (*LR*) as a function of various parameters, anode to cathode separation (*d*), field emitter sharpness (*S*) determined by the ellipsoid radii ratio R_1/R_2, and ellipsoid parameter (R_1) using their self-focusing effect in the NFESEM configuration and taking into account some geometrical input from our experiments. For instance, the geometry of the field emitters used in the simulation are obtained by fitting the experimental TEM images of actual tips used for NFESEM. Accordingly, the polycrystalline W-tip shown in Fig. 14 was used to test the application of the self-focusing method. An ellipsoid was fitted to exclude the surface contaminants and include only the pure W-field emitter. The figure was also used to determine a geometrical approximation of the apex for a spherical surface. This was used for the image potential-energy calculation as well as to estimate r_{phys}.

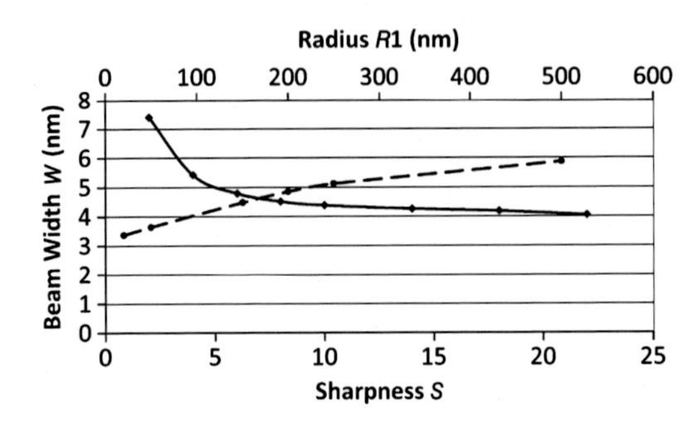

Figure 20 Lateral resolution of the simulated electron beam in the conventional SEM mode ($d >> l$: total length of conducting field emitter) as a function of (A) the field emitter sharpness S at a constant ellipsoid parameter $R_1 = 170$ nm (lower horizontal axis – associated curve: solid line) and (B) ellipsoid parameter R_1 at a constant field emitter sharpness $S = 10$ (upper horizontal axis – associated curve: dashed line). Reprinted from Ultramicroscopy, Vol. 125, Kyritsakis, A. & Xanthakis, J. P., Beam spot diameter of the near-field scanning electron microscopy, 24–28, Copyright 2013, with permission from Elsevier [OR APPLICABLE SOCIETY COPYRIGHT OWNER].

The simulated *LR* with respect to field emitter sharpness S and ellipsoid parameter R_1 are shown in Fig. 20 for parameters associated with conventional SEM, *i.e.* $d >> l$: total length of conducting field emitter and $R_1 = 170$ nm. In this simulation, the *LR* was determined by the Full width at half maximum (FWHM) of the angular distribution of the current density. The local electric field F was set at 5 V/nm, and the FWHM was calculated at a distance of 10 nm. Accordingly, the simulation establishes a strong dependence between *LR* on S for S values ranging from 2–20; where *LR* decreases by almost a factor of two. However, increasing the sharpness beyond this range will not significantly reduce the *LR*. The *LR* is similarly sensitive to the ellipsoid radius R_1, which exhibits a similar decrease in *LR* for decreasing R_1. Note that the S was held constant at a value of 10 and $R_2 = 17$ nm for the *LR*–R_1 simulation.

Most important to this study is *LR* as a function of d for NFESEM operating conditions, *i.e.* 10 nm $< d <$ 100 nm, and $F = 5$ V/nm. The three curves in Fig. 21 represent the different ellipsoid fitting parameters for Fig. 14 at three different scan areas, see Fig. 1 in Kyritsakis and Xanthakis (2013); where the associated fitting parameters are $R_1 = 170$ nm and $R_2 = 17$ nm (solid line), $R_1 = 650$ nm and $R_2 = 30$ nm (dashed line), and $R_1 = 100$ nm and $R_2 = 20$ nm (dotted line). The curves show fairly linear behavior and the *LR* is consistently smaller than d. In fact, the *LR* is always less than half of the distance from the field emitter apex to the anode for all of the fitting parameters associated with Fig. 14.

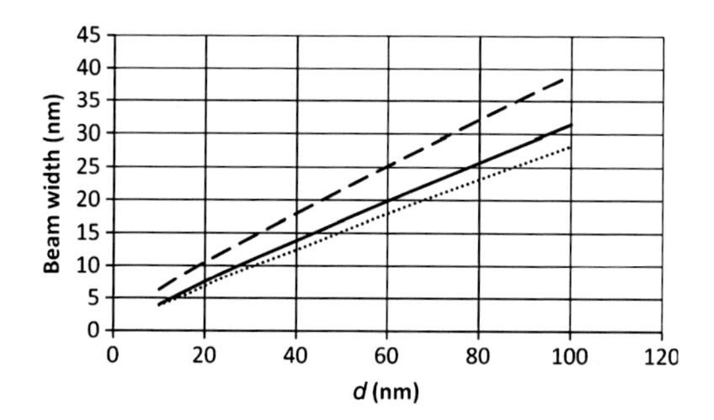

Figure 21 Lateral resolution of the simulated electron beam in the NFESEM mode (10 nm < d < 100 nm) as a function of (A) anode to cathode separation distance d for $R_1 = 170$ nm and $R_2 = 17$ nm (solid line), $R_1 = 650$ nm and $R_2 = 30$ nm (dashed line), and $R_1 = 100$ nm and $R_2 = 20$ nm (dotted line). Reprinted from Ultramicroscopy, Vol. 125, Kyritsakis, A. & Xanthakis, J. P., Beam spot diameter of the near-field scanning electron microscopy, 24–28, Copyright 2013, with permission from Elsevier [OR APPLICABLE SOCIETY COPYRIGHT OWNER].

An additional analysis of the V_a–d experimental curves has been performed using a version of the CERN Minuit minimizing procedure based on FORTRAN language that allows fitting of an arbitrary number of points with any type of function. The script was implemented in order to investigate the theory explained in Section 3.1, in which the system is only characterized by the geometry of the field emitter and not the dynamic of the FE. The fitting function used to minimize the V_a–d curve has been defined as:

$$d = \begin{cases} F_0 \dfrac{d}{r_{tip}} & (d < R_0) \\ F_0 \cdot C \cdot r_{tip} \cdot \ln\left(\dfrac{N \cdot d}{r_{tip}}\right) + V_0 & (d > R_0) \end{cases} ; \qquad (4.13)$$

where F_0 is the local electric field, which is fixed in CC-mode, C is a prefactor (from the theory $C = 1/2$), r_{tip} is the expected effective radius, r_{eff}, of the field emitter, N is a form parameter similar to a factor depending on the geometry (2 for parabolic, 4 for hyperbolic shape) and V_0 is a constant, which allows for continuity in the changing of regimes – from linear to logarithmic – that are defined from the point R_0. This assumes that the planar anode is located at the focal plane of the hyperboloid (see Kirk, 2010b), which has been applied to many contemporary FE models (see Zuber et al., 2002). In order to optimize the fitting procedure, the points near R_0 should be neglected. Since it would be difficult to estimate a proper value

Table 1 Example of fitting parameters used for a V_a–d curve.

#	Name	Starting value	Standard deviation	Lower limit	Upper limit	units
1	R_0	2.1d1	1.d-1	5.0	1.d3	nm
2	R	3.0d1	1.d-2	0.	0.	nm
3	F	2.0d0	1.d-1	0.	0.	V/nm
4	Xi_m	2.0d1	0.d-1	0.	0.	V
5	Xi_M	4.3d1	0.d-1	0.	0.	V

R is the effective radius; whereas the X_i-values are the upper and lower limit of the points that are not fitted.

Table 2 Actual fitting parameters used for a V_a–d curve.

Measure	R_0	R [R_{eff}]	F	V_0	Xi_m	Xi_M
a[1]	5.86	5.81	2.77	10.58	20.00	43.00
b[1]	8.15	5.65	2.61	13.49	30.00	45.00
e[1]	7.81	6.38	2.62	12.97	42.85	198.98
f[1]	5.01	9.10	2.02	9.24	0.43	45.00
a[2]	8.86	5.81	2.76	5.00	20.00	43.00

[1] Parabolic approximation ($N = 2$).
[2] Hyperbolic approximation ($N = 4$).

for R_0, it has been decided to let this parameter to be free, defining two regions of points used for the fitting procedure. This region was chosen arbitrarily with respect to the voltage. Therefore the independent variables that have to be minimized by Minuit are in Table 1.

These variables are defined in a file named: *parameter.inp*. It is then clear that if the standard deviation for an arbitrary variable is zero, the variable is considered to be a constant (*i.e.* already minimized). Conversely, when the lower limit and the upper limits are set to zero, the variable is free to vary. It is beneficial to define a domain, in which the variable is defined; in order to avoid the variable assuming unreasonable values.

In order to improve the statistics of the simulation we tried to vary both: 1) the set parameters and 2) the different factors discussed in Table 1, *viz*; R_0, R, F, Xi_m and Xi_M. Table 2 shows the results of this type of statistical alterations. Another limitation of these fitting parameters is the inability to distinguish the parabolic from the hyperbolic approximation, *i.e.* changing the argument of the logarithm by a factor of two, as this does not significantly alter the fitting curve. Subsequently, the fitting results with the hyperbolic approximation yields:

$$V(d) = 2.76\ \text{V/nm} \cdot \frac{5.81\ \text{nm}}{2} \cdot \ln\left(\frac{4 \cdot d}{5.81\ \text{nm}}\right) + 5; \qquad (4.14)$$

where a reasonable value of the local field $F = 2.76$ V/nm and $r_{\text{eff}} = 5.81$ nm was deduced.

More recently, there has been another interpretation of the V_a–d measurements that do not consider contemporary FE theory. In particular Prof. Pescia (ETHZ) et al., see Cabrera et al. (2013), have modeled the field emitter-plane interface using a conical, metallic tip positioned a distance d away from an infinite plane at a constant potential V_a. This model departs from the convention of having the planar anode being located at the focal plane of the field emitter. They have found that when characteristic I_{FE} curves are measured as a function of two independent variables; namely, V_a and d, the family of curves can be collapsed into one single scaling curve. This scaling curve is I_{FE} as a function of the scaling variable $V_a d^{-\lambda}$; where λ is constant with a value of approximately 0.22 for d ranging from 10^{-9} m to 10^{-3} m. This is an important result, as it considers the effect of the emitter shank on the local electric field at the apex. Moreover, Cabrera et al. have simulated the FE emitter with a series of small spheres aligned along the direction perpendicular to the planar anode and with increasing radius of curvature. The terminating sphere has a radius of 4 nm and an apex electric field of 4 V/nm. Their results show that the barrier width increases for decreasing radii as d increases, which is indicative of non-planar FE (see Sections 4.2 and 4.3).

A similar result can be deduced using contemporary FE theory, which was originally proposed by Dr. R. Forbes.[5] Consider the "physically complete" F–N-type equation Eq. (3.6) at the reference maximum current density, J_0:

$$J_0 = \left(\lambda_Z a\varphi^{-1} F_0^2\right) P_F \exp\left[-\nu_F b\varphi^{3/2}/F_0\right]. \tag{4.15}$$

Instead of relating the local, reference electric field, F_0, with the voltage-to-local-electric-field conversion factor and β_0, it will be related via the equation:

$$F_0 = \frac{V_a}{\zeta_0}; \tag{4.16}$$

where ζ_0 is the reference conversion length that not only considers the apex, but the shank as well. Accordingly, the FE current through the refer-

[5] "Comments on the voltage scaling of field electron emission current–voltage characteristics" presented at the 27th International Vacuum Nanoelectronics Conference, held in Engelberg, Switzerland, 6–10 July, 2014.

ence emission area, A_0, is:

$$I_0 = A_0 \left(\lambda_Z a\varphi^{-1}\zeta^{-2} V_a^2\right) P_F \exp\left[-\nu_F b\varphi^{3/2}\zeta_0 / V_a\right]. \tag{4.17}$$

It is evident that the emission current is going to be dominated by the V_a/ζ_0 term. Moreover, R. Forbes has shown that experimental scaling parameter $R(d)$, defined by Cabrera et al., can be predicted in the general form by:

$$R(d) = \frac{V_a(d_0)}{V_a(d)} = \frac{\zeta_0(d_0)}{\zeta_0(d)}. \tag{4.18}$$

This assumes that the local field is the same for any tip-plane distance d. In order to advance this theory, the local work function of the apex surface must be considered, because this will vary the local field producing "patch" fields. Nevertheless, it follows that the experimental scaling parameter is the ratio of the conversion lengths.

Prof. J. Xanthakis has extended this scaling relationship to include varying tip radii (see Kyritsakis, Xanthakis, & Pescia, 2014). Xanthakis et al. observe consistent scaling when $d >> r_{tip}$; however, there is a deviation when $d \sim r_{tip}$. The deviation is associated with an increased angular dependence of the apex electric field with d. Consequently, the experimental scaling parameter is better defined by the ratio of the local electric fields. In accordance with the ETHZ study, Xanthakis et al. can clearly define three regions of varied d-dependence for V_a–d curves:

1. $d << r_{tip}$: parallel-plate arrangement of the conversion factor β discussed in Section 3.1;
2. $r_{tip} << d << l$ (total length of conducting emitter): a power law is observed, contrary to the aforementioned logarithmic dependence;
3. $d >> l$: the field emitter is distant from anode, so the local electric field does not vary with small variation of d. The scaling function is also a constant.

This work led Xanthakis et al. to derive a physically complete F–N-type equation with a correction factor for non-planar field emitters (Kyritsakis & Xanthakis, 2015).

5. TOPOGRAPHIC IMAGING

Using the calibration of the PMT HAMAMATSU R 268 (see Appendix A) it is now possible to **quantitatively** analyze most of the topographic measurements performed with the same SED. To this end a

Matlab-based software, which can be also used to analyze future measurements, was developed.

The measurements are first visualized with the WSxM program from Nanotec Electrónica (Horcas et al., 2007), of which in addition to the image visualization of the measured FE current and the PMT voltage measurement, allows for the processing of the images. Nevertheless, in order to execute a systematic processing of the images, the raw data of the selected measurements are saved in a text file using a matrix form, which can be used as default in the WSxM. The matrices are then loaded from the Matlab function which, in combination with the calculated calibration, is able to transform the signal of the PMT back to the input signal of the SED.

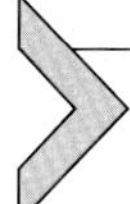

6. ALTERNATIVE CONTRAST MECHANISMS

6.1 Magnetic

The interaction of electrons with the sample has already been discussed in Kirk (2010a). However, when the target contains ferromagnetically ordered spins, the exchange interaction between the spin of the incoming electrons and the spin of the target electrons must be taken into account. There are two experimental situations in which the exchange interaction produces measurable effects.

In the first instance, the primary electron beam is spin-polarized. There are a number of methods used to generate spin-polarized FE (see Getzlaff, 2009); however, the focus of this study was iron coated tungsten (Fe–W) field emitters. Here, it is the spin dependence of the surface potential that is of interest. This spin dependence occurs because in a ferromagnet, the density of the majority electrons differs from that of the minority electrons. It follows that the exchange interaction between a given tunneling electron and the other electrons depends on its spin. Accordingly, the surface of a ferromagnet acts as a spin filter with respect to electrons incident on it and are subsequently field-emitted into vacuum. We have fabricated emitters, in which a tungsten FE tip was prepared via the aforementioned procedure and subsequent layers of iron were deposited on the tungsten surface. Our ability to control the surface magnetization of these type of field emitters is shown in Fig. 22. Here, an external magnetic field, 8 kA/m, is applied along the axis of a tip. Then the tip was imaged using SEMPA (Ramsperger, 1996); where, the black and white bands display

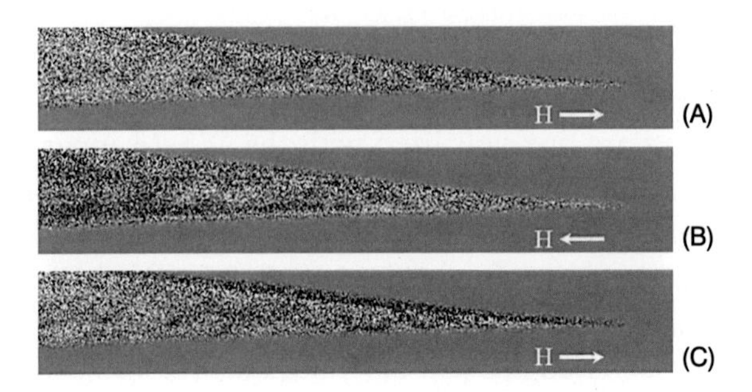

Figure 22 10 k magnification SEMPA images of the magnetization vector. The arrows indicate the direction of the magnetizing field. These SEMPA micrographs demonstrate the reversal of the magnetization vector parallel to the tip axis, after applying an 8 kA/m *H*-field. (A) in-plane with external magnetizing field, *H*, applied parallel to tip axis (B) in-plane anti-parallel to tip axis (C) reversed back to the in state (A) Taken from Kirk et al. (2010).

the spin asymmetry of the surface states. Similar tips were used to generate a spin-polarized beam of the electrons that subsequently interacts with a magnetized sample. This technique was implemented in order to enhance the spin dependent scattering at the sample surface by employing field emitted electrons with a spin asymmetry. In such an experiment, the intensity of the backscattered electrons is detected. The intensity of both elastically and inelastically backscattered electrons depends on the relative orientation of the spin of the incident electrons and the spin of the target electrons, via exchange interaction determining a spin dependence of the elastic and inelastic mean free path. This results in an asymmetry of the scattered intensity when the spin polarization of the incident beam is oriented parallel or antiparallel to the sample surface spin states. The asymmetry is particularly strong at low energies (see Passek et al., 1996), which was the range our FE measurements. Some preliminary results using this mode are presented later in this section.

In a second experiment, one uses an unpolarized primary beam of electrons. Both elastically and secondary electrons are polarized via the exchange interaction of the primary electron spins with the spin polarized sample spins. The polarization of the electrons allows us, similar to the asymmetry in the previous mode, to map the magnetic contrast of the target. One application of this second mode, SEMPA (Koike & Hayakawa, 1984), uses a remote unpolarized electron source to perform magnetic microscopy with SEs in an energy range of ≤ 5 eV. The implementation of this experiment with a localized electron source is being developed[6] as a follow up to this work.

In order to demonstrate the prospect of magnetic contrast imaging, a W (110) substrate was prepared using the method discussed in Kirk (2010a). Upon allowing ample time for substrate cooling, 30–60 minutes, an ultrathin film of Fe has been subsequently grown via UHV molecular beam epitaxy (MBE, deposition rate $\sim$ 0.2 ML/minute). The thicknesses of the sample ranged from 1.1–3.2 MLs. The heteroepitaxial growth of Fe on W (110) – in this specific thickness range – exhibits a variety of magnetic features, due to stress on the Fe islands from the lattice mismatch with the substrate (Przybylski, Kaufmann, & Gradmann, 1989). As a result, Fe grows pseudomorphically on W (110) at room temperature for the first ML (Sander, Skomski, Schmidthals, Enders, & Kirschner, 1996).

The quality of our films was verified by Auger electron spectroscopy and low energy electron diffraction. Typically the layer height is 2.25 Å, as confirmed by line profiles made by STM. We observe island growth perpendicular to step edge for a thickness of 1.8 ML. This STM image will later be used to characterize NFESEM features.

After growth, the samples were transported to the scanning Kerr microscope chamber for magnetic characterization. Longitudinal magneto-optical Kerr effect (MOKE) detected a square hysteresis loop for a magnetic field, greater than 16 kA/m, applied along the easy in-plane direction, *i.e.* along the $\langle 1, -1, 0 \rangle$ direction. No hysteresis was detected for samples with a thickness less than $\sim$ 2 MLs. The hysteresis loop in Fig. 23 shows typical easy-axis behavior for the $\sim$ 3 MLs thick film of Fe.

Both NFESEM and STM were performed on selected Fe-coated samples. In general, the samples were prepared and measured on the same day, to reduce surface contamination that significantly reduce and/or alter the SE yield. The topographic imaging will now be discussed, emphasizing the comparison between NFESEM and STM. It is also important to note that the FE current (I_{FE}) and the probing electron beam energy (E_P) are analogous to the tunneling current (I_T) and bias voltage (V_B) used in STM.

Although major features of the surface can be observed in the raw NFESEM images, the images were enhanced to reveal more contrast by correcting for the tilt of the sample (procedure detailed in Kirk, 2010a). Two separate samples were prepared each having 3.2 MLs of Fe and one

[6] This endeavor is currently being funded by a National Science Foundation: Major Research Instrumentation development grant NSF-1531997.

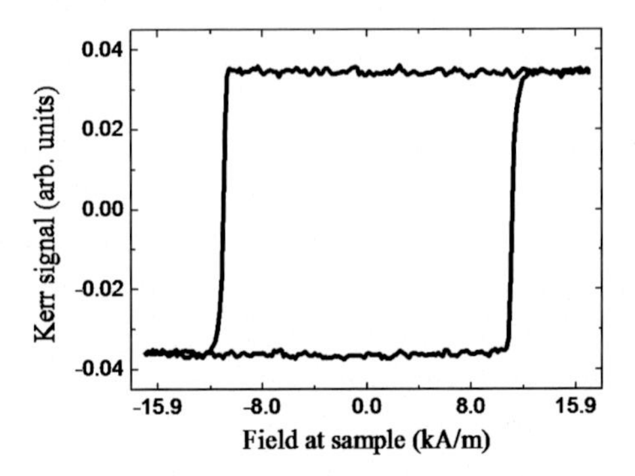

Figure 23 Hysteresis loop of a $\sim$ 3 ML film of Fe on W (110) obtained with a spatially resolved MOKE at the center of the sample. *The data in this figure are published with permission of ETH Zürich.*

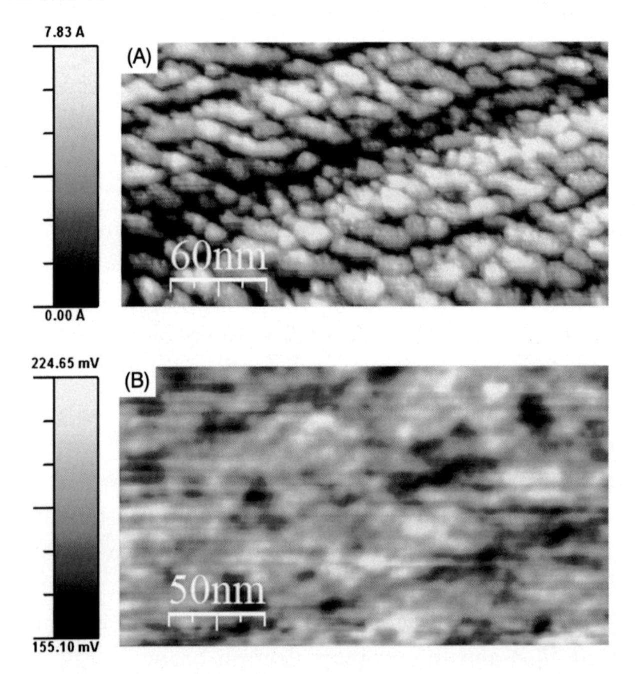

Figure 24 Micrographs of 3.2 MLs of Fe on a W (110) substrate imaged with (A) STM ($I_T = 0.15$ nA and $V_B = 0.2$ V) and (B) NFESEM ($d = 30$ nm, $I_{FE} = 54$ nA and $E_p = 42$ eV) (Kirk et al., 2010).

was imaged with STM, Fig. 24A, and the other with NFESEM, Fig. 24B. In both images, we recognize at least four levels of contrast, indicating the presence of three Fe layers on top of a – partially visible – W (110) substrate (the darkest regions). Although these samples were prepared on

Table 3 SNFESEM assessment of spin-dependent scattering asymmetry (SSA) for 1.8 MLs and 3.2 MLs of iron on a W (110) substrate.

Film Thickness (MLs)	Layer Height (Å)	SED signal (mV)	*SSA*[1]
1.8 MLs	2.25 ± 0.25	7 ± 1	N/A
3.2 MLs	2.25 ± 0.25	14 ± 2	33.3%

[1] This assumes that signal variations between the two samples are mainly due to their magnetic properties.

different days the surface structures bear striking similarities, *e.g.* the terrace width. Note the large range in NFESEM vertical intensity, as compared to the STM scale bar. A rough estimate shows that the SE signal varies approximately 9 mV/Å, which is due to the sensitivity of the FE current via to small deviations of the electric field as the field emitter is scanned over the surface. There are a number of horizontal "streaks" visible in Fig. 24B, which are most-likely caused by instabilities in the FE current originating from adsorbate motion between the tip and the sample. Potentially, a reduction in the scanning speed, from 330 nm/second, would reduce this effect.

The growth of the Fe islands, normal to the step edge – as observed by STM in Fig. 24A – is also visible using NFESEM imaging (see Fig. 24B). Some lateral features of the iron islands, much smaller than the tip-to-sample separation of 60 nm, can be observed in the NFESEM figure. Moreover, the signal from the SED varies significantly, even though vertical topographic features are only a few Ångstrom.

Assuming that there is no surface contaminants, *e.g.* iron-oxide, to alter the SE yield, the amount of detected electrons per layer should remain constant. This was confirmed in the NFESEM visualization, and the SE yield will be discussed in terms of the SED signal (mV)/thickness (Å). Fig. 24 was generated using a polycrystalline tungsten emitter that has been coated with Fe and a magnetic field was applied in the same direction as the sample, in-plane. Since the sample magnetization cannot be reversed at the present SNFESEM set-up, the results of the thick iron sample will be compared to the relevant features of the thin iron SNFESEM image. In practice the SE yield of the layers are mainly dependent on the beam current and energy as well as the tip-sample separation; therefore one must compare similar values to make conclusions about the SE yield. The results are tabulated in the following Table 3; where the SSA is calculated as the relative polarization between a magnetic sample and a non-magnetic sample. That is to say if the magnetization were reversed, the SSA should be twice as large.

Obviously the effective SE yield in the thicker sample is greater than the measured SE yield in the thin sample, even though the primary beam current was less. This could possibly be due to the background from the tungsten substrate that could have reduced the detected SE yield from the iron or the SE emission of the thicker iron sample is simply higher. The measured line profiles of the thin sample exhibit no enhancement or reduction in SE yield when the first layer is compared to the second layer. Another relevant question is whether the mean free path (MFP) increases at lower energies; thus generating more SE that can be detected by the SED. In other words, a contrast inversion could be observed, meaning that the parallel alignment of field emitted electron spins and the surface magnetization vector would generate more detected secondary electrons. However, this increase in SE yield would primarily be due to the chemical nature of the material and not the magnetic state. Although this is a possibility, and there is not sufficient evidence to entirely refute this argument, previous measurements show no sharp increase in SE yield at lower energies. In fact, the SE yield of a single material is usually significantly less at low primary beam energies, *e.g.* 20 eV, even at high FE currents. Moreover, the SE yield is only increased when both the tip and the sample are magnetized. Still there is no observation of non-topographic variations in the NFESEM image, *i.e.* domains, which are expected to be normal to the step edges, in the $\langle 1, -1, 0 \rangle$-direction. The scan range was also limited to 1 μm by 1 μm, so multiple domains are not likely to be found. However, previous studies in this group suggest that the ultrathin iron films are in a single domain state (Ramsperger, 1996). Final confirmation of the magnetic domain state of this sample would require additional magnetic surface imaging, *e.g.* magnetic force microscopy.

More recently, the second NFESEM at ETHZ was modified for high resolution topographic imaging. Accordingly, this requires the tip-sample separation to be reduced in a controlled manner, as well as the ability to maintain a stable FE current. The group has prepared a non-magnetic sample of 0.2 MLs of Fe a top of a clean W (110) substrate, shown in Fig. 25. They have successfully imaged the surface using NFESEM (see Zanin et al., 2014a) at a tip-sample separation of 11 nm, 10 nm, and 9 nm, which demonstrates a higher scan stability *cf.* the previous prototype. A subsequent STM measurement, in exactly the same scan area, was used as a reference for comparison. It is evident that the two images in Fig. 25 bear a striking resemblance; however the contrast is inverted. This means that the Fe islands that decorate the W (110) terraces appear as indentations (Fig. 25A)

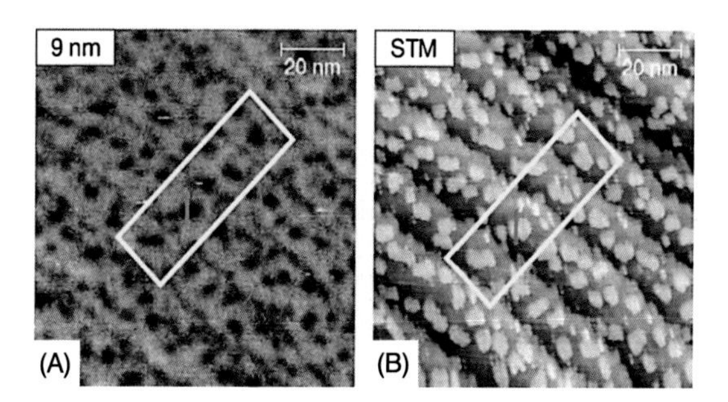

Figure 25 100 nm × 100 nm micrographs of 0.2 MLs of Fe on a W (110) substrate, imaged with (A) NFESEM ($d = 9$ nm, $I_{FE} = 300$ nA and $E_P = 24$–35 eV) and (B) STM ($I_T = 0.07$ nA and $V_B = 0.2$ V). The yellow rectangle is used to highlight groups of Fe islands on the W (110) terraces. Note that the Fe islands appear as indentation in the NFESEM image; whereas they appear as protrusions in the STM micrograph. Taken from Zanin et al. (2014a). © 2014 IEEE. Reprinted, with permission, from IEEE Proceedings, Improving the topografiner technology down to nanometer spatial resolution.

rather than protrusions on the surface (Fig. 25B). This is a similar result reported by First et al. (see Section 2.4.1 and First et al., 1991) for an array of rectangular Permalloy bits on a Si substrate. Although the primary beam of electrons used by First et al. (Fig. 3B) and the second ETHZ prototype (Fig. 25A) is slightly less than the one used by the first prototype in Fig. 24, the apparent unforeseen variations in SE yield could be due to a low electron energy contrast inversion. The researchers at ETHZ have also observed that there was an increase in contrast – in lieu of a reduction of detected SEs – as the tip-sample separation was reduced (Zanin et al., 2014a). Accordingly, the chemical nature of the material(s) and contrast inversion must be taken into consideration when examining the magnetic properties of samples. This further stresses the importance of very low energy electron interactions, which greatly broaden our understanding of materials.

6.2 Chemical

Despite an interest in low energy electrons to characterize materials, the understanding of the electron transport, electron energy loss processes and generation of SEs is poorly understood. This often presents difficulties in the interpretation – as demonstrated in the previous section – of the results that these techniques provide. Low electron beam energies may also allow for additional contrast mechanisms, *e.g.* crystalline, doping, and enhanced topographic (see Cazaux, 2012). The MFP of electrons is expected to increase with decreasing energy at very low energies which suggests the interesting possibility of enhanced penetration[7] of very slow electrons through thin films. The field emitted electrons interact via electron–electron excitation at very long energies; whereas higher primary electron energies can excite plasmons. The penetration depth will vary in accordance with this transition. The NFESEM generally operates at energies below the excitation energies of bulk plasmons, so the associated incident wavelength is approaching values greater than the inter-atomic spacing. The sample reflectance decreases at these lower energies (Cazaux, 2012), which increases the SE yield.

In accordance with the analysis of the SEs and the investigation of additional contrast mechanisms, an energy analyzer has been constructed during D. Zanin's doctoral work at ETHZ with expert Prof. M. Erbudak. This energy analyzer was fabricated in consideration of the deflection of the ejected electrons by the strong electric field between the tip and the sample, similar to the aforementioned EELS device using localized field emission electron scanning microscopy. The spectrometer could acquire both plasmon loss and SE information simultaneously which will considerably assist in the identification and characterization of nanoscale particles and as such complements the recent developments of the NFESEM. Preliminary measurements of a cleaved sample of GaAs (110) for two different tip-sample separations, 10 nm and 110 nm. The results are shown in Fig. 26, and the two spectra were normalized to the energy of the field emitted electrons incident on the sample surface. Both spectra exhibit inelastic and elastic electron excitation, confirming occurrences of energy loss. The spectra show a dependence of the tip-sample separation; where more SEs are generated at a greater separation. However, the most remarkable result is that more SEs are generated than BSEs for primary electron beam energies around 35 eV at a tip-sample separation of 110 nm. This is of great importance because low energy SEs, $E_P < 5$ eV, exhibit the highest spin asymmetry (Koike & Hayakawa, 1984). It follows that a low primary energy beam is suitable for polarization analysis, which would have presented a major challenge otherwise.

It is known that the elastic reflectance of very slow electrons, namely below about 20 to 30 eV where the inelastic scattering becomes sufficiently weak, is inversely proportional to the local density of (unoccupied) electron states coupled to the incident electron beam. The possibility of direct imaging of the local density of states is of great importance for the solid state

[7] These penetration depth variations are to be distinguished from those generated via magnetic interactions, which also vary the SE yield.

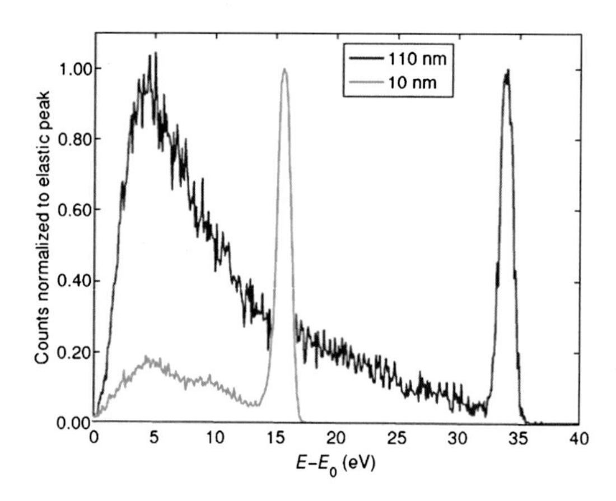

Figure 26 Electron energy spectra of electrons detected from a cleaved GaAs (110) surface. The spectra were recorded at a tip-sample separation of 10 nm and 110 nm, and the spectra are both normalized to their respective primary field emitted electron beam energies. Taken from Zanin et al. (2014b). © 2014 IEEE. Reprinted, with permission, from IEEE Proceedings, Detecting the topographic, chemical, and magnetic contrast at surfaces with nm spatial resolution.

physics in general but also for all branches of micro- and nano-electronics. This is one future outlook for these type of experiments.

6.3 Future Prospects

There is currently a large, international collaboration directed towards a greater understanding of low energy (1–200 eV) electron interactions, Marie Curie Initial Training Network (FP7): Simdalee2. The purpose of this network is to provide researchers with state-of-the-art, and beyond, technical and physical "know-how" on advanced nanoscale characterization procedures. The NFESEM will play a central role in this endeavor. The need of high spatial and temporal resolution analysis is significant in modern sciences, *e.g.* toward spintronic and novel semiconductor-based devices. Optimizing scanning electron/probe and transmission electron-type instruments means addressing the main processes, *i.e.* electron generation, electron-specimen interaction and detection of the subsequent, ejected electrons. The network consists of industrial and academic partners, each with a strong background in one or more of the aforementioned topics; thus bridging the gap between these disciplines by constructively merging them together. This network is primarily about the characterization of surfaces or near surface regions with low energy electrons and they will pursue the following goals: (1) optimizing the beam size by correlating contemporary FE theory with high resolution holographic measurements of magnetic and electric fields of FE tips with different shapes, both with and without primary electron optics; (2) putting the understanding of the contrast mechanism of electron beam techniques on a sound footing by comparing physical models with novel benchmark spectra acquired using a coincidence technique; (3) improving detection as well as understanding of emitted energy-, angular-, and spin-dependent spectra. This issue will be addressed for the common case of detectors in the a field-free environment within the ultrafast temporal regime, and for the special case when the emitted electrons encounter an electric field prior to detection; (4) ultimately, progress in the aforementioned fields will lead to the development of an innovative prototypical methodology for nanoscale characterization with electron beams.

7. CONCLUSIONS

In this work, we have shown that STM operating in the constant height, FE *modus operandi* can generate a well-defined primary electron beam, when positioned in close proximity to a conducting surface. There are essentially three substantial results that must be highlighted:

1. At high FE currents, tens of nano-amperes, detectable variations in the FE current – **on the order of 20%** – can be used to detect features of atomic height. This is a result of the localized electron excitation used in NFESEM, which is mainly an electrostatic effect originating from the geometry of the field emitter and the specimen surface. These preliminary images show that the SE yield is strongly dependent on the vertical displacement, since the variations of the FE current produce the bulk of the SE signal.
2. As part of the development of field emitters, we have fabricated a number of polycrystalline tungsten tips. These emitters have been imaged with both SEM and TEM, which allows us to determine the tip shape with high spatial resolution. These field emitters exhibit characteristic electrostatic behavior which is directly correlated to the field emitter geometry and chemical makeup. We are now able to discern a relationship between the electron microscopy images and the characteristic curves measured with the field emitters. The measurement is performed by approaching the sample in constant current mode to within one nanometer of the sample surface. A fixed FE current

means that the electric field at the tip surface, used to extract the electrons, is also effectively constant. The bias voltage to the sample is then increased. The resultant curve is characteristic of the electric field behavior between the tip and the sample; in accordance with the NFESEM geometry. Our electrostatic measurements indicate a "self-focusing" effect, which is a lensless converging of the primary electron beam. This is a major accomplishment that contributes to the explanation of the high resolution capabilities of the NFESEM.

3. The latest frontier of NFESEM research is alternate contrast mechanism imaging. We are currently pursuing efforts in both magnetic and chemical contrast imaging. Although the two could be considered mutually exclusive, they become intertwined at low primary beam energies. In particular, contrast inversions have been observed for energies under 40 eV. Nevertheless, we believe that our electron energy loss spectroscopy studies will shed light on this phenomenon. We have also observed large SE backgrounds for primary electron energies as low as 35 eV on a GaAs sample, which is a promising result for future studies of the spin polarization analysis for magnetic contrast imaging.

APPENDIX A. DETECTOR CALIBRATION

This appendix describes the different methods used for the calibration of the detector. By calibration we mean the determination of the ratio between the intensity of electrons entering the secondary electron detector (SED) and the primary field emission (FE) intensity, or $f_{SE}\delta$. This is a very important parameter also for future developments, as it provides some clue about the actual amount of electrons that finally escape the surface and reach the detector. For calibration purposes, we replace the local field emitter by a beam electrons from a remote electron gun (EG).

We have used a thermionic LEG62 electron gun. Here, a tungsten (W) filament was heated with a current of a couple of amperes ($I_{fil} = 1$–3 A), generating thermal electron emission. Electrons were then accelerated towards an anode with a voltage of some kVs, which produces an emission current of hundreds of microamperes ($I_{emis} = 100$–200 μA). A successive focusing system was needed to confine the beam, because of the high spread out of the electrons. In the case of the LEG62 it is possible to confine the beam using an extractor, a condenser, and a final focus. Two methods are used to measure the beam current.

Both methods have been developed in an *ad hoc* UHV chamber, completely separate from the NFESEM setup. All the calibrations were made using a vacuum between a low range UHV and high range High Vacuum (HV), since it was not possible to recreate the same extreme UHV conditions ($p \approx 2 \cdot 10^{-11}$ mbar) existing in the actual chamber. The detector calibration is measured in a base pressure between $p \approx 10^{-8}$–10^{-9} mbar, which is between 2–3 orders of magnitude higher than the system used for NFESEM. This inevitably leads to a slightly different behavior of the electrons and, accordingly, of the SED. The main procedure for the SED calibration entails measuring the current entering the SED with high precision, *i.e.* the current that arrives at the scintillator, and comparing this value to the signal coming out of the SED. In order to improve the statistics and perform a repeatable calibration, each setup has been tested manually as well as with the help of a computerized controller, which has simultaneously controlled the electron gun and measured the output signals.

Part I: Faraday Cup Measurement

This first method is based on the assumption that it is possible to measure the signal at the entrance of the SED by replacing the detector with a metallic target, which will measure the impinging electrons. On the one hand, the electron gun accelerates the electrons up to some kilo-electron volts (keV)s; therefore, a high positive potential, to attract the electrons, is not needed. On the other hand, a Faraday cage, affixed to the metallic target, will trap the back scattered and the secondary electrons (SE)s generated by the impinging of the high energy electrons coming from the EG. Once the current is measured, the Faraday cage will be retracted and replaced by the SED, and the measurement is repeated. Fig. 27 shows a schematic picture of this setup. In this setup, the substitution is possible because the motion axes of the two targets – the Faraday Cage and the Scintillator of the SED – are perpendicular to each other with a wide displacement range. It is important to isolate the measurement in order to not introduce unexpected external noise, for instance, when applying a high voltage to the titanium ring of the SED.

The procedure of this measurement can therefore be divided in two steps:

- Input current measurement at the Faraday Cage
- Output current measurement with the Photomultiplier

Since the two measurement systems are physically separated, after each displacement it is important to properly re-align the target with the electron

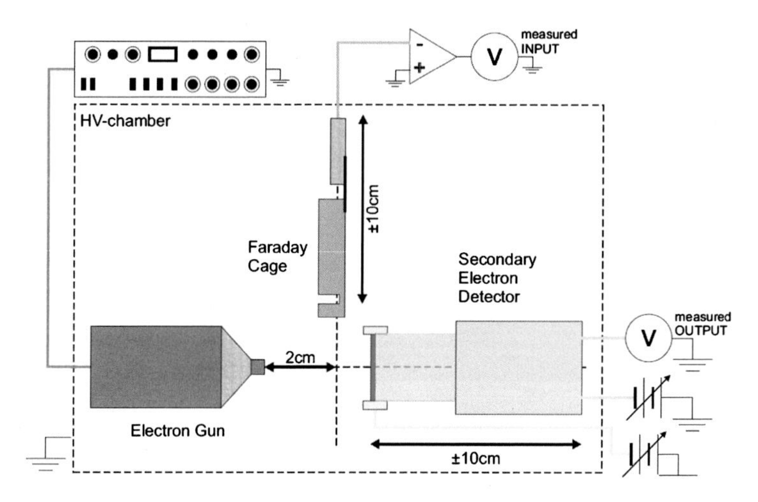

Figure 27 Sketch of the setup used for the first calibration with the Faraday Cage. The calibration with the Faraday cage was done in a small vacuum system at pressure of $p \approx 10^{-9}$ mbar, *i.e.* under HV conditions. The PMT used was a HAMAMATSU R 268.

source. It is therefore essential to execute a whole curve for the input current with respect to a repeatable characteristic of the EG (*e.g.* the emission current) and then reproduce the measurement for the output current with the SED.

This mechanical replacement has some drawbacks. Indeed, between the Faraday Cage measurement and the SED measurement the condition (*e.g.* pressure, temperature, gun stability, *etc.*) may change. A possible solution to reduce this problem would be to exchange the Faraday Cage with the scintillator of the SED, after each single point of a measurement. However this increases the duration of the measurement and also introduces a misalignment between two points of the measurement. Even if the Faraday cage would trap every SE, a possible improvement of this system can be to introduce a bias voltage of between 50 and 100 V in order to attract and trap more electrons and thus refine the measurement of the input current.

Once the input current is defined, two other parameters must be specified for the aim of calibration: the voltage applied to the PMT (V_{PMT}) – needed to convert the incoming light into a measurable current – and the accelerating voltage (V_{ACC}) – applied at the titanium ring at the front end of the SED in order to convert the input electrons into light. The calibration is then complete once the gain factor – defined as the ratio between

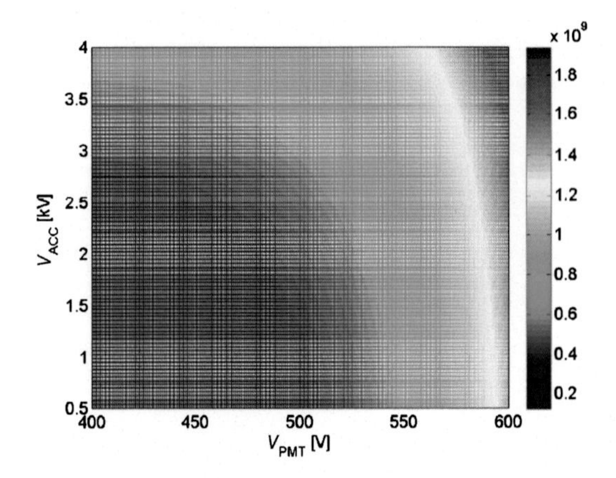

Figure 28 Gain versus V_{PMT} and V_{ACC} for the PMT Hamamatsu R 268. The color-map on the right represents the total gain in V/A, as determined by Eq. (A.1). While imaging, the output voltage is detected. Using such a reference calibration, the input current entering the detector can be determined, and related to the probe current, to obtain the actual SE yield for the current entering the detector (Zanin et al., 2012).

the output signal in volts (V) and the input signal in amperes (A) has been established, for different V_{PMT} and V_{ACC}.

For fixed V_{ACC} and V_{PMT}, the output SED voltage depends linearly on input current, and the SED gain can be determined from the slope. Fig. 28 shows the gain as a function of V_{PMT} and V_{ACC} within a color-coded two-dimensional plot.

The total gain function $f_{TOT\text{-}gain}$ can be approximated phenomenologically as:

$$\begin{aligned} f_{TOT\text{-}gain} &= f_{gain}(V_{PMT}) + f_{gain}(V_{ACC}) \\ &= \exp(0.01831 \cdot V_{PMT}) + 2.251 \cdot 10^4 \\ &\quad + \exp(0.5489 \cdot V_{ACC}) 6.752 \cdot 10^7. \end{aligned} \tag{A.1}$$

This formula reveals an exponential behavior of the output signal both when the applied potential of the PMT is varied, and when the acceleration voltage is changed.

Part II: Complementary Measurement of the Input Current

The duration and reproducibility problems connected with the Faraday cup measurement of the input current led us to an alternative method of gain determination. This consists of completely removing the Faraday Cage

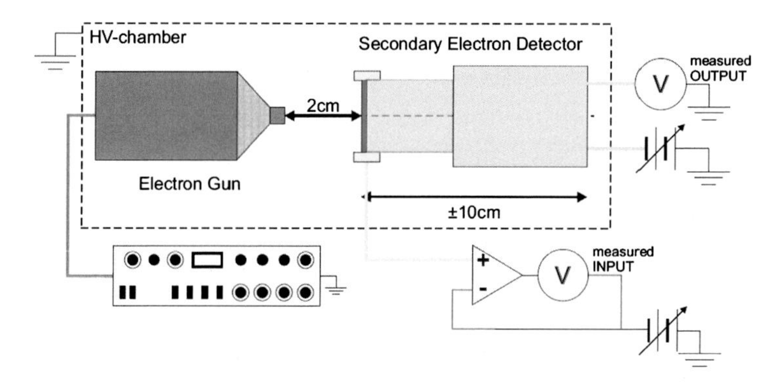

Figure 29 Sketch of the setup used for the calibration method without a Faraday cup.

and taking advantage of the conductivity of the scintillator material. The current entering the detector is measured directly at the entrance of the SED allowing for the simultaneous measurement of the input and output currents. The electrons impinging on the scintillator surface are trapped by the high bias of the titanium ring (3 kV), before going towards the ground the current is amplified which allows for small currents to be measured in the presence of a high voltage. A sketch of the system used for this calibration is shown in Fig. 29.

This solution significantly reduces the duration of the measurement and thus of the calibration, without the need of any assumption for measuring the input current. In addition, the simultaneous measurement of the input and output currents allows for the calibration to occur under well defined, constant experimental conditions. Fig. 30 shows the gain as a function of V_{PMT} and V_{ACC} within a color coded two-dimensional plot.

Note that by using this set-up we were able to determine the dependence of the calibration curves on the actual, relative position between the EG and the SED. It turned out that no detectable dependence was observed. We however point out that the angular spread of the gun and of the "real" SEs appearing in NFESEM could be strongly different. In this second case, we cannot exclude some position dependence.

APPENDIX B. COMPARISON OF r_{eff} VS. r_{phys}

Table 4 shows the estimated tip radii and errors. The FE data for the tips marked with a "*b*" were obtained after SEM imaging and a second annealing procedure, *i.e.* with the build-up layer. The tips that were annealed

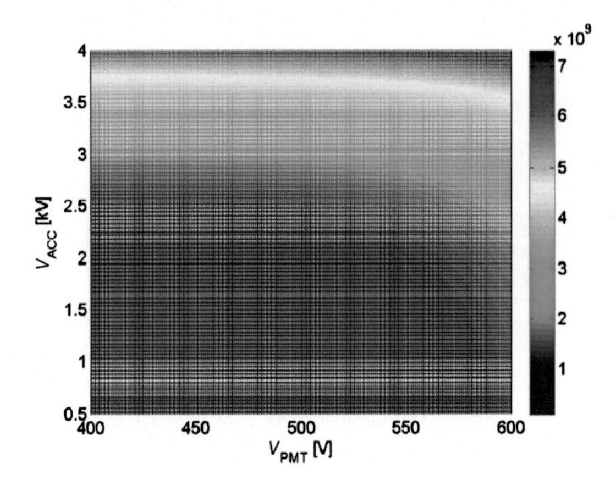

Figure 30 Gain surface for the ET Enterprises 9924B. The color-map on the right represents the total gain in V/A (Zanin et al., 2012).

Table 4 Summary of field emitter data for fabrication optimization.

Tip number	Effective emission radius (nm)	Error (nm)	Radius from images (nm)
A 2	18.70	±4.68	18
A 3	12.97	±3.24	23
A 3b	21.03	±5.26	–
A 4	11.25	±2.81	–
A 5	16.84	±4.21	14
A 5b	21.03	±5.26	N/A
A 6	9.93	±2.48	7
A 7 w\ O_2	11.26	±3.70	–
A 9 w\ O_2	9.26	±2.39	10.5
A 10 w\ O_2	7.65	±1.99	7.5
M 1 w\ O_2	9.11	±2.41	20
M 2 w\ O_2	no I_{FE}	N/A	57
M 3	13.37	±3.47	N/A
M 3 w\ O_2	12.72	±3.33	11.5
M 5	9.65	±2.47	N/A
M 5 w\ O_2	no I_{FE}	N/A	16
M 7	7.11	±1.97	N/A
M 7 w\ O_2	no I_{FE}	N/A	N/A
M 8 w\ O_2	7.10	±1.92	N/A

Stockklauser (2010). *Reprinted with permission from Anna Stockklauser.*

in oxygen atmosphere are denoted by "w\O_2." The symbol "–" means that the tip radius could not be extracted from the SEM images due to contamination. For some tips field emission data was recorded before and after oxygen annealing. Tips M 5 and M 7 exhibited "good" field emission before oxygen annealing but did not show any afterwards. This occurrence was unexpected and there is currently no explanation. Tip M 3 showed the expected behavior, which is discussed in detail in Section 4.1.2. The tips M 7 and 8 could not be imaged under the SEM because they were damaged after FE characterization.

Altogether tips A 9 (w\$O_2$ treatment), A 10 w\$O_2$, and M 3 w\$O_2$ provide the best, most complete and comparable results. Theory predicts that the effective emitter radius is smaller than the physical. When the error margin is taken into account, the expected correlation between the physical and the effective emitter radius (see Section 4.1.1) can be observed for these three tips.

ACKNOWLEDGMENTS

This project was led by Dr. Taryl L. Kirk, who has been appointed a visiting assistant professorship at Rowan University and later at the College of New Jersey. Dr. Richard Forbes and Prof. John Xanthakis have both provided extensive support on FE theory. Prof. Xanthakis has developed the self-focusing effect based on our experimental results and has been instrumental in determining the lateral resolution of NFESEM. We hope to eventually measure this behavior directly via the electron holography technique of Prof. Rafal E. Dunin-Borkowski and Dr. Takeshi Kasama. The imaging of all of our electron sources was made at the microscopy center of ETHZ (ScopeM formerly known as EMEZ), EMPA, and DTU. We would like to thank Elisabeth Müller, Peter Weber, and Karsten Kunze of EMEZ for their expertise. The high resolution TEM image of the electron source apex Fig. 14 was provided by Magdalena Parlinska-Wojtan at the Center for Electron Microscopy, EMPA. We have also had fruitful discussions with Prof. Jacques Cazaux (University Reims), Dr. Chris Walker (University of York), and Prof. Wolfgang Werner (TU Vienna) about the interactions of electrons with matter. Furthermore, we would like to acknowledge the work of Prof. Richard Palmer (University of Birmingham) and Prof. Juan Sáenz (Universidad Autónoma de Madrid) that has enabled us to construct the NFESEM and determine the resolution capabilities. The support by Swiss National Foundation (SNF), the Swiss Federal Innovation Promotion Agency (CTI), and the ETH Zürich is gratefully acknowledged. Dr. Kirk has recently acquired a National Science Foundation (NSF): Major Research Instrumentation grant dedicated to the construction of a high resolution NFESEM with polarization analysis, *i.e.* NFESEMPA. This work will be supported by the grant NSF-1531997. Dr. Kirk is now working at Educational Testing Service; however, he has returned to Rowan University as a visiting assistant professor to construct the microscope in collaboration with Rutgers University and Villanova University.

FIGURE ACKNOWLEDGMENTS

Reprinted from Advances in Imaging and Electron Physics, Vol. 170, Zanin, D. A., Cabrera, H., De Pietro, L. G., Pikulski, M., Goldman, M., Ramsperger, U., Pescia, D., & Xanthakis, J. P., Chapter 5 – Fundamental Aspects of Near-Field Emission Scanning Electron Microscopy, 227–258, Copyright 2012, with permission from Elsevier [OR APPLICABLE SOCIETY COPYRIGHT OWNER].

Reprinted from Ultramicroscopy, Vol. 125, Kyritsakis, A. & Xanthakis, J. P., Beam spot diameter of the near-field scanning electron microscopy, 24–28, Copyright 2013, with permission from Elsevier [OR APPLICABLE SOCIETY COPYRIGHT OWNER].

DISCLAIMER

The opinions expressed in this document are that of the author and not of Prof. Danilo Pescia and his Microstructure Research group.

Any opinions expressed in the publication are those of the author and not necessarily of Educational Testing Service.

REFERENCES

Allenspach, R. (1994). The attraction of spin-polarized SEM. *Physics World*, *7*, 44–49.

Allenspach, R., & Bischof, A. (1989). Spin-polarized secondary electrons from a scanning tunneling microscope in field emission mode. *Applied Physics Letters*, *54*(6), 587–589.

Bauer, E., Duden, T., & Zdyb, R. (2002). Spin-polarized low energy electron microscopy of ferromagnetic thin films. *Journal of Physics. D, Applied Physics*, *35*, 2327–2331.

Binnig, G., & Rohrer, H. (1982). Surface studies by scanning tunneling microscopy. *Physical Review Letters*, *49*, 57–61.

Cabrera, H., Zanin, D. A., De Pietro, L. G., Michaels, Th., Thalmann, P., Ramsperger, U., ... Abanov, Ar. (2013). Scale invariance of a diodelike tunnel junction. *Physical Review. B, Condensed Matter and Materials Physics*, *87*, 115436.

Cazaux, J. (2010). Material contrast in SEM: Fermi energy and work function effects. *Ultramicroscopy*, *110*, 242–253.

Cazaux, J. (2012). Reflectivity of very low energy electrons (< 10 eV) from solid surfaces: Physical and instrumental aspects. *Journal of Applied Physics*, *111*, 064903.

Cho, B., Ichimura, T., Shimizu, R., & Oschima, C. (2004). Quantitative evaluation of spatial coherence of the electron beam from low temperature field emitters. *Physical Review Letters*, *92*, 246103.

Cutler, P. H., He, J., Miskowsky, N. M., Sullivan, T. E., & Weiss, B. (1993). Theory of electron emission in high fields from atomically sharp emitters: Validity of the Fowler–Nordheim equation. *Journal of Vacuum Science & Technology. B, Microelectronics and Nanometer Structures Processing, Measurement and Phenomena*, *11*, 387–391.

Edgcombe, C. J. (2010a). New dimensions for field emission: Effects of structure in the emitting surface. *Advances in Imaging and Electron Physics*, *162*, 77–127.

Edgcombe, C. J. (2010b). The transverse structure of cold field electron emission. *Ultramicroscopy*, *110*, 1454–1459.

Edgcombe, C. J., & De Jonge, N. (2007). Deduction of work function of carbon nanotube field emitter by use of curved-surface theory. *Journal of Physics. D, Applied Physics*, *40*, 4123–4128.

Erying, C. F., Mackeown, S. S., & Millikan, R. A. (1928). Fields currents from points. *Physical Review*, *31*, 900–909.

Eves, B. J., Festy, F., Svensson, K., & Palmer, R. E. (2000). Scanning probe energy loss spectroscopy: Angular resolved measurements on silicon and graphite surface. *Applied Physics Letters*, 77, 4223–4225.

Festy, F., Svensson, K., Laitenberger, P., & Palmer, R. E. (2001). Imaging surfaces with reflected electrons from a field emission scanning tunnelling microscope: Image contrast mechanisms. *Journal of Physics. D, Applied Physics*, *34*, 1849–1852.

Fink, H.-W. (1986). Mono-atomic tips for scanning tunneling microscopy. *IBM Journal of Research and Development*, *38*, 260–263.

First, P. N., Stroscio, J. A., Pierce, D. T., Dragoset, R. A., & Celotta, R. J. (1991). A system for the study of magnetic materials and magnetic imaging with the scanning tunneling microscope. *Journal of Vacuum Science & Technology. B, Microelectronics and Nanometer Structures Processing, Measurement and Phenomena*, *9*(2), 531–536.

Forbes, R. G. (2006). Simple good approximations for the special elliptic functions in standard Fowler–Nordheim tunneling theory for a Schottky–Nordheim barrier. *Applied Physics Letters*, *89*, 113122.

Forbes, R. G., & Deane, J. H. B. (2010). Comparison of approximations for the principal Schottky–Nordheim barrier function $v(f)$, and comments on Fowler–Nordheim plots. *Journal of Vacuum Science & Technology. B, Microelectronics and Nanometer Structures Processing, Measurement and Phenomena*, *25*(2), C2A33–C2A42.

Fowler, R. H., & Nordheim, L. (1928). Electron emission in intense electric fields. *Proceedings of the Royal Society of London. Series A*, *119*(781), 173–181.

Fursey, G. (2005). *Field emission in vacuum microelectronics*. New York: Kluwer/Plenum.

Getzlaff, M. (2009). *Spin-polarized scanning tunneling microscopy*. Weinheim: Wiley-VCH Verlag GmbH & Co. KGaA.

Goldstein, J., Newbury, D. E., Joy, D. C., Lyman, C. E., Echlin, P., Lifshin, E., . . . Michael, J. R. (2003). *Scanning electron microscopy and X-ray microanalysis* (3rd ed.). New York: Plenum Press.

He, J., Cutler, P. H., & Miskovsky, N. M. (1991). Generalization of Fowler–Nordheim field emission theory for non-planar metal emitters. *Applied Physics Letters*, *59*, 1644–1646.

Horcas, I., Fernández, R., Gómez-Rodríguez, J. M., Colchero, J., Gómez-Herrero, J., & Baro, A. M. (2007). WSXM: A software for scanning probe microscopy and a tool for nanotechnology. *Review of Scientific Instruments*, *78*, 013705.

Joy, D. C., & Jakubovics, J. P. (1968). Direct observation of magnetic domains by scanning electron microscopy. *Philosophical Magazine*, *17*(145), 61–69.

Kirk, T. L. (2010a). Near field emission scanning electron microscopy. *Applied Electron Microscopy* [Angewandte Elektronenmikroskopie] *Vol. 9*. Berlin: Logos Verlag.

Kirk, T. L. (2010b). Near field emission scanning electron microscopy. In *Microscopy: Science, technology, applications and education*. Badajoz: Formatex.

Kirk, T. L., De Pietro, L. G., Cabrera, H. L., Bähler, T., Maier, U., Ramsperger, U., & Pescia, D. (2010). *Ultra-high resolution microscopy via localized electron excitations* (Poster presentation. International Microscopy Congress 2010, Rio de Janeiro, Brazil, September 20–24, 2010).

Kirk, T. L., De Pietro, L. G., Pescia, D., & Ramsperger, U. (2009). Electron beam confinement and image contrast enhancement in near field emission scanning electron microscopy. *Ultramicroscopy*, *109*, 463–466.

Kirk, T. L., Ramsperger, U., & Pescia, D. (2009). Near field emission scanning electron microscopy. *Journal of Vacuum Science & Technology. B, Microelectronics and Nanometer Structures Processing, Measurement and Phenomena*, *27*(1), 152–155.

Kirk, T. L., Scholder, O., De Pietro, L. G., Ramsperger, U., & Pescia, D. (2009). Evidence of nonplanar field emission via secondary electron detection in near field emission scanning electron microscopy. *Applied Physics Letters*, *94*(15), 153502.

Koike, K., & Hayakawa, K. (1984). Scanning electron microscope observation of magnetic domains using spin-polarized secondary electrons. *Japanese Journal of Applied Physics*, *23*, L187–L188.

Kyritsakis, A., Kokkorakis, G. C., Xanthakis, J. P., Kirk, T. L., & Pescia, D. (2010). Self-focusing of field emitted electrons at an ellipsoidal tip. *Applied Physics Letters*, *97*(2), 023104.

Kyritsakis, A., & Xanthakis, J. P. (2013). Beam spot diameter of the near-field scanning electron microscopy. *Ultramicroscopy*, *125*, 24–28.

Kyritsakis, A., & Xanthakis, J. P. (2015). Derivation of a generalized Fowler–Nordheim equation for nanoscopic field-emitters. *Proceedings of the Royal Society A*, *471*, 20140811.

Kyritsakis, A., Xanthakis, J. P., Kirk, T. L., & Pescia, D. P. (2011). Lateral resolution of the NFESE microscopy and the existence of self focusing of electrons. In *International vacuum nanoelectronics conference proceedings*.

Kyritsakis, A., Xanthakis, J. P., & Pescia, D. P. (2014). Scaling properties of a non-Fowler–Nordheim tunnelling junction. *Proceedings of the Royal Society A*, *470*, 20130795.

Melmed, A. J. (1991). The art and science and other aspects of making sharp tips. *Journal of Vacuum Science & Technology. B, Microelectronics and Nanometer Structures Processing, Measurement and Phenomena*, *9*(2), 601–608.

Midgley, P. A., & Dunin-Borkowski, R. E. (2009). Electron tomography and holography in materials science. *Nature Materials*, *8*, 271–280.

Müllerová, I., & Frank, L. (2007). Very low energy scanning electron. In *Microscopy: Science, technology, applications and education*. Badajoz: Formatex.

Passek, F., Donath, M., & Ertl, K. (1996). Spin-dependent electron attenuation lengths and influence on spectroscopic data. *Journal of Magnetism and Magnetic Materials*, *159*, 103–108.

Philibert, J., & Trixier, R. (1969). Effets de contraste cristallin en microscopie électronique à balayage. *Micron*, *1*, 174–186.

Pierce, D. T. (1988). Spin-polarized electron microscopy. *Physica Scripta*, *38*, 291–296.

Przybylski, M., Kaufmann, I., & Gradmann, U. (1989). Mössbauer analysis of ultrathin ferromagnetic Fe(110) films on W(110) coated by Ag. *Physical Review. B, Condensed Matter and Materials Physics*, *40*, 8631–8640.

Ramsperger, U. (1996). *Structural and magnetic investigations of ultrathin microstructures* (Ph. D. thesis). Zurich: Swiss Federal Institute of Technology.

Sáenz, J. J., & García, R. (1994). Near field emission scanning electron tunneling microscopy. *Applied Physics Letters*, *65*(23), 3022–3024.

Sakurai, T., & Müller, E. W. (1973). Field calibration using the energy distribution of field ionization. *Physical Review Letters*, *30*, 532–535.

Sakurai, T., & Müller, E. W. (1977). Field calibration using the energy distribution of free-space field ionization. *Journal of Applied Physics*, *48*, 2618–2625.

Sander, D., Skomski, R., Schmidthals, C., Enders, A., & Kirschner, J. (1996). Film stress and domain wall pinning in sesquilayer iron films on W (110). *Physical Review Letters*, *77*, 2566–2569.

Scholder, O. (2009). *Optimization of spin-polarized field emission in near field emission electron microscopy* (M. Sc. thesis). Zurich: Swiss Federal Institute of Technology.

Stockklauser, A. (2010). *Nanoscale characterization of the field emitter for near field emission scanning electron microscopy* (Semesterarbeit). Zurich: Swiss Federal Institute of Technology.

Tondare, V. N., van Druten, N. J., Hagen, C. W., & Kruit, P. (2003). Stable field emission from W tips in poor vacuum conditions. *Journal of Vacuum Science & Technology. B, Microelectronics and Nanometer Structures Processing, Measurement and Phenomena, 21*(4), 1602–1606.

Vida, G., Josepovits, V. K., Gyor, M., & Deak, P. (2003). Characterization of Tungsten surfaces by simultaneous work function and secondary electron emission measurements. *Microscopy and Microanalysis, 9*, 337–342.

Wiesendanger, R. (1994). *Scanning probe microscopy and spectroscopy: Methods and application.* Cambridge: Cambridge University Press.

Young, R., Ward, J., & Scire, F. (1972). An instrument for measuring surface microtopography. *Review of Scientific Instruments, 43*(7), 999–1011.

Zanin, D. A., Cabrera, H., De Pietro, L. G., Pikulski, M., Goldman, M., Ramsperger, U., ... Xanthakis, J. P. (2012). Chapter 5 – Fundamental aspects of near-field emission scanning electron microscopy. *Advances in Imaging and Electron Physics, 170*, 227–258.

Zanin, D. A., Erbudak, M., De Pietro, L. G., Cabrera, H., Vindigni, A., Pescia, D., & Ramsperger, U. (2014a). Improving the topografiner technology down to nanometer spatial resolution. In *27th international vacuum nanoelectronics conference proceeding*.

Zanin, D. A., Erbudak, M., De Pietro, L. G., Cabrera, H., Vindigni, A., Pescia, D., & Ramsperger, U. (2014b). Detecting the topographic, chemical, and magnetic contrast at surfaces with nm spatial resolution. In *27th international vacuum nanoelectronics conference proceeding*.

Zuber, J. D., Jensen, K. L., & Sullivan, T. E. (2002). An analytical solution for microtip field emission current and effective emission area. *Journal of Applied Physics, 91*, 9379–9384.

CHAPTER THREE

Nonscalar Mathematical Morphology

Jasper van de Gronde[1], Jos B.T.M. Roerdink
University of Groningen, Groningen, The Netherlands
[1]Corresponding author: e-mail address: jasper.vandegronde@gmail.com

Contents

1. INTRODUCTION

Mathematical morphology has traditionally mostly been used on binary and grayscale images. These kinds of images share a common theme: they admit a very natural total order on the value space. The theory behind mathematical morphology is equipped to deal with more general situations

Advances in Imaging and Electron Physics, Volume 204
ISSN 1076-5670
https://doi.org/10.1016/bs.aiep.2017.09.004

through the use of lattices, but this has only been used sparingly. It is reasonable to say that the main reason has been that lattices are ill-suited to many such situations. In particular, we will consider both higher dimensional data, as well as categorical data.

When we deal with multidimensional data, like vector-valued data, we usually have some ideas about what kinds of operations "should not matter". For example, if $\rho(a) = b$ for some vector-valued images a and b, and a' is another vector-valued image such that the vectors at all positions are rotated by some fixed rotation – $a'(x) = r(a(x))$ – then we might expect $\rho(a')(x)$ to simply be a rotated version of $\rho(a)(x)$ as well. It can be shown that this type of constraint is incompatible with imposing a lattice structure on the vector space, at least if we assume the lattice is compatible with the vector space in a particular (and sensible) way. Section 3 shows how one can work around this by representing vectors using a frame rather than a basis. Unfortunately, this means using a larger and/or more complicated representation, and the answers you get might not correspond to the frame-based representation of any particular vector, leading to a possible loss of information, as well as a loss of the properties that usually characterize morphological operators. This motivates the introduction of sponges in Section 4: a generalization of lattices, sponges are easier to adapt to multidimensional spaces, and we present several examples of sponges. Interestingly, we can show that sponges allow us to recover at least some parts of the usual morphological theory based on lattices.

Categorical data presents another interesting challenge: there is no up or down, no large or small. While a vector at least has a magnitude, different categories are just different. We cannot even use something like a median algebra, since we also do not have a concept of "betweenness": different categories are just different. Section 5 shows how the concept of n-ary morphology can be used to deal with this case. We also briefly touch upon data where each value is not a single category, but a vector of likelihoods for each category.

Finally, in Section 6 we very briefly touch upon some other ideas people have used to apply morphology to non-traditional data, and when they should be preferred over the approaches detailed in the earlier sections. But first, we will have a closer look at the traditional morphological framework.

Code for all of the examples in this work can be found at http://bit.ly/2u95Z8W.

2. LATTICE-BASED MATHEMATICAL MORPHOLOGY

In its original form, mathematical morphology is a set-theoretical approach to image analysis (Matheron, 1975; Serra, 1982). The key idea is to define image transformations based on *shape* information. In the simplest case of binary images, small subsets, called *structuring elements*, of various forms and sizes are translated over the image plane to perform shape extraction. Such morphological image transformations have, on the one hand, an intuitive geometrical interpretation, and on the other hand can be precisely formulated as algebraic image operators. In contrast to traditional linear image processing based on concepts such as convolution and frequency analysis, the morphological image operators focus on the *geometrical* content of images and are *nonlinear*.

The mathematical description of morphological image operators has been extensively developed within the framework of complete lattice theory (Serra, 1988; Heijmans, 1994; Heijmans & Ronse, 1989; Ronse & Heijmans, 1991). Also, the case of vector-valued data, such as color or hyper-spectral images, has been addressed. An important tool in morphological image analysis is the algebraic decomposition and synthesis of image operators in terms of elementary operations. Once such an algebraic decomposition is available, it enables efficient implementations on digital computers, see e.g. (Giardina & Dougherty, 1988; Soille, 2003).

In the remainder of this section we summarize the most important concepts of lattice-based morphology, starting with the case of binary images and then considering gray-scale images. We pay attention to the concept of *invariance* of lattice operators, leading to group morphology (Roerdink, 2000), which has been the inspiration for looking at the frame-based approach discussed in later sections. There are many other lattice-based morphological operators we will not discuss here, such as adaptive filters (Cheng & Venetsanopoulos, 2000; Maragos & Vachier, 2009; Ćurić, Landström, Thurley, & Hendriks, 2014; Lerallut, Decencière, & Meyer, 2007; Roerdink, 2009), morphology for color and vector images (Aptoula & Lefèvre, 2007b; Ledoux & Richard, 2016; Angulo, 2007; Velasco-Forero & Angulo, 2011a), and matrix fields (Burgeth, Papenberg, Bruhn, Welk, Feddern, & Weickert, 2005; Burgeth, Bruhn, Papenberg, Welk, & Weickert, 2007; Burgeth & Kleefeld, 2017).

2.1 Morphology for Binary Images

Let E be the Euclidean space $\mathbb{R}^n$ or the discrete grid $\mathbb{Z}^n$. By $\mathcal{P}(E)$ we denote the set of all subsets of E ordered by set-inclusion. A binary image can be represented as a subset X of E. The space E is a commutative group under vector addition: we write $x + y$ for the sum of two vectors x and y, and $-x$ for the inverse of x. The following two algebraic operations are fundamental for mathematical morphology of binary images:

$$\textit{Minkowski addition}: X \oplus A = \{x + a : x \in X, a \in A\} = \bigcup_{a \in A} X_a = \bigcup_{x \in X} A_x$$

$$\textit{Minkowski subtraction}: X \ominus A = \bigcap_{a \in A} X_{-a},$$

where $X_a = \{x + a : x \in X\}$ is the translate of the set X along the vector a.

Let the *reflected* or *symmetric* set of A be denoted by $\check{A} = \{-a : a \in A\}$. The transformations $\delta_A : \mathcal{P}(E) \to \mathcal{P}(E)$ and $\varepsilon_A : \mathcal{P}(E) \to \mathcal{P}(E)$ defined by

$$\delta_A(X) := X \oplus A = \{h \in E : (\check{A})_h \cap X \neq \emptyset\} \tag{1}$$

$$\varepsilon_A(X) := X \ominus A = \{h \in E : A_h \subseteq X\}, \tag{2}$$

are called *dilation* and *erosion* by the structuring element A, respectively.

There exists a *duality relation* with respect to set-complementation (X^c denotes the complement of the set X): $X \oplus A = (X^c \ominus \check{A})^c$, i.e. dilating an image by A gives the same result as eroding the background by $\check{A}$. To any mapping $\psi : \mathcal{P}(E) \to \mathcal{P}(E)$ we associate the *dual* mapping $\psi' : \mathcal{P}(E) \to \mathcal{P}(E)$ by

$$\psi'(X) = \{\psi(X^c)\}^c. \tag{3}$$

By composition of dilation and erosion, other important transformations can be built; in particular, the *opening* α_A, which is an erosion following by a dilation, and the *closing* ϕ_A, which is a dilation followed by an erosion:

$$\alpha_A(X) := \delta_A(\varepsilon_A(X)) = \bigcup\{A_h : h \in E, A_h \subseteq X\}$$

$$\phi_A(X) := \varepsilon_A(\delta_A(X)) = \bigcap\{(\check{A}^c)_h : h \in E, (\check{A}^c)_h \supseteq X\}.$$

Also opening and closing have an intuitive geometric interpretation, as can be observed from the above formulas. The opening of X is the union of all the translates of the structuring element which are included in X. The closing of X by A is the complement of the opening of X^c by $\check{A}$.

The dilation δ_A, erosion ε_A, opening α_A, and closing ϕ_A are all examples of *translation-invariant* binary image operators. For any operator $\psi : \mathcal{P}(E) \to \mathcal{P}(E)$ we say it is translation-invariant when $\psi(X_h) = (\psi(X))_h$ for all $h \in E$.

2.2 Complete Lattice Framework

We now summarize the main concepts from lattice theory needed in this paper, cf. Heijmans (1994), Serra (1988). For a general introduction to lattice theory, see Birkhoff (1961).

A *complete lattice* $(\mathcal{L}, \leq)$ is a partially ordered set $\mathcal{L}$ with order relation $\leq$, a supremum or join operation written $\bigvee$, and an infimum or meet operation written $\bigwedge$, such that every (finite or infinite) subset of $\mathcal{L}$ has a supremum (smallest upper bound) and an infimum (greatest lower bound). In particular there exist two universal bounds, the least element written $O_{\mathcal{L}}$ and the greatest element $I_{\mathcal{L}}$. Barring such a least and greatest element, one can still have a *conditionally complete* lattice, in this case every subset of $\mathcal{L}$ that is bounded from below has a meet, and every subset that is bounded from above has a join.

For the case of binary images, the relevant lattice is the power lattice $\mathcal{P}(E)$ of all subsets of the set E, the order relation is set-inclusion $\subseteq$, the supremum is the union $\bigcup$ of sets, the infimum is the intersection $\bigcap$ of sets, the least element is the empty set $\emptyset$, and the greatest element is the set E itself. The power lattice $\mathcal{P}(E)$ is an atomic complete Boolean lattice, and conversely any atomic complete Boolean lattice has this form.

Mappings

The composition of two mappings ψ_1 and ψ_2 on a complete lattice $\mathcal{L}$ is written $\psi_1\psi_2$ or $\psi_1 \circ \psi_2$, and instead of $\psi\psi$ we also write ψ^2. An *automorphism* of $\mathcal{L}$ is a bijection $\psi : \mathcal{L} \to \mathcal{L}$ such that for any $a, b \in \mathcal{L}$, $a \leq b$ if and only if $\psi(a) \leq \psi(b)$. If ψ_1 and ψ_2 are operators on $\mathcal{L}$, we write $\psi_1 \leq \psi_2$ to denote that $\psi_1(a) \leq \psi_2(a)$ for all $a \in \mathcal{L}$.

A mapping $\psi : \mathcal{L} \to \mathcal{L}$ is called:

- *idempotent*, if $\psi^2 = \psi$;
- *extensive*, if for every $a \in \mathcal{L}$, $\psi(a) \geq a$;
- *anti-extensive*, if for every $a \in \mathcal{L}$, $\psi(a) \leq a$;
- *increasing (isotone, order-preserving)*, if $a \leq b$ implies that $\psi(a) \leq \psi(b)$ for all $a, b \in \mathcal{L}$;

- a *closing*, if it is increasing, extensive and idempotent;
- an *opening*, if it is increasing, anti-extensive and idempotent.

Let $\mathcal{L}$ and $\mathcal{M}$ be complete lattices. A mapping $\psi : \mathcal{L} \to \mathcal{M}$ is called:

- a *dilation*, if $\psi(\bigvee_{i\in I} a_i) = \bigvee_{i\in I} \psi(a_i)$ and $\psi(O_{\mathcal{L}}) = O_{\mathcal{M}}$;
- an *erosion*, if $\psi(\bigwedge_{i\in I} a_i) = \bigwedge_{i\in I} \psi(a_i)$ and $\psi(I_{\mathcal{L}}) = I_{\mathcal{M}}$.

Let $\varepsilon : \mathcal{L} \to \mathcal{M}$ and $\delta : \mathcal{M} \to \mathcal{L}$ be two mappings, where $\mathcal{L}$ and $\mathcal{M}$ are complete lattices. Then the pair (ε, δ) is called an *adjunction between $\mathcal{L}$ and $\mathcal{M}$*, if for every $a \in \mathcal{M}$ and $b \in \mathcal{L}$, the following equivalence holds: $\delta(a) \leq b \iff a \leq \varepsilon(b)$. If $\mathcal{M}$ coincides with $\mathcal{L}$ we speak of an *adjunction on $\mathcal{L}$*. It has been shown (Gierz, Hofmann, Keimel, Lawson, Mislove, & Scott, 1980; Heijmans & Ronse, 1989; Ronse & Heijmans, 1991) that in an adjunction (ε, δ), ε is an erosion and δ a dilation. Also, for every dilation $\delta : \mathcal{M} \to \mathcal{L}$ there is a unique erosion $\varepsilon : \mathcal{L} \to \mathcal{M}$ such that (ε, δ) is an adjunction between $\mathcal{L}$ and $\mathcal{M}$; ε is given by $\varepsilon(b) = \bigvee\{a \in \mathcal{M} : \delta(a) \leq b\}$, and is called the *upper adjoint* of δ. Similarly, for every erosion $\varepsilon : \mathcal{L} \to \mathcal{M}$ there is a unique dilation $\delta : \mathcal{M} \to \mathcal{L}$ such that (ε, δ) is an adjunction between $\mathcal{L}$ and $\mathcal{M}$; δ is given by $\delta(a) = \bigwedge\{b \in \mathcal{L} : a \leq \varepsilon(b)\}$, and is called the *lower adjoint* of ε. Finally, for any adjunction (ε, δ), the mapping $\delta\varepsilon$ is an opening on $\mathcal{L}$ and $\varepsilon\delta$ is a closing on $\mathcal{M}$. In the case that $\mathcal{L}$ and $\mathcal{M}$ are identical, one sometimes refers to such openings and closings as *morphological* or *adjunctional* (Heijmans, 1994).

As a specific case we may consider the lattice of gray scale functions. Let $\mathcal{L} = \mathrm{Fun}(E, \mathcal{T})$ denote the complete lattice of gray scale functions with domain E, whose range is a complete lattice $\mathcal{T}$ of gray values. Then the following gray scale dilation-erosion transform pair on $\mathcal{L}$ can be defined ($f \in \mathrm{Fun}(E, \mathcal{T})$, $b \in \mathrm{Fun}(B, \mathcal{T})$, $B \subseteq E$):

$$\delta_{b,B}(f)(x) = \bigvee_{y\in B}(f(x-y)+b(y)), \quad \varepsilon_{b,B}(f)(x) = \bigwedge_{y\in B}(f(x+y)-b(y)), \quad \forall x \in E.$$

Gray scale opening and closing are defined by composition, as in the binary case.

Invariance

Invariance of image operators can be captured by group actions (Robinson, 1982; Suzuki, 1982). When $\mathbb{T}$ is an (automorphism) group of two lattices $\mathcal{L}$ and $\mathcal{M}$, a mapping $\psi : \mathcal{L} \to \mathcal{M}$ is called $\mathbb{T}$-*invariant* if it commutes with all $\tau \in \mathbb{T}$, i.e., if $\psi(\tau(a)) = \tau(\psi(a))$ for all $a \in \mathcal{L}$, $\tau \in \mathbb{T}$. More generally, a mapping $\psi : \mathcal{L} \to \mathcal{M}$, with $\mathcal{L}$ and $\mathcal{M}$ two distinct lattices, is said to be invariant to $\mathbb{T}$ if $\psi(\tau(a)) = \rho_\tau(\psi(a))$ for all $a \in \mathcal{L}$, $\tau \in \mathbb{T}$ and some representation $\rho : \mathbb{T} \to \mathbb{S}$ of $\mathbb{T}$ on $\mathcal{M}$, where $\mathbb{S}$ is a group on $\mathcal{M}$.

For the case of binary images, the symmetry group is the Euclidean translation group. For other situations, different groups may be relevant. For example, some images have radial instead of translation symmetry (Serra, 1982, p. 17), requiring a polar group structure. The generalization of morphology for binary images with arbitrary *abelian* symmetry groups was worked out by Heijmans (1987), see also Roerdink and Heijmans (1988); for the general lattice case, see Heijmans and Ronse (1989), Ronse and Heijmans (1991). The further generalization to lattices with a non-abelian symmetry group was made by Roerdink (1993, 2000).

The general setup is to start from a group $\mathbb{T}$ acting *transitively* on the image space E. We say that $\mathbb{T}$ is *transitive on* E if for each $x, y \in E$ there is a $g \in \mathbb{T}$ such that $gx = y$, and *simply transitive* when this element g is unique. A *homogeneous space* is a pair $(\mathbb{T}, E)$ where $\mathbb{T}$ is a group acting transitively on E. Any transitive abelian group $\mathbb{T}$ is simply transitive. For the Boolean case, the object space of interest is again the lattice $\mathcal{P}(E)$ of all subsets of E. The general strategy is to make use of the results for the simply transitive case (corresponding to the situation where $\mathbb{T}$ is the Euclidean translation group), by 'lifting' subsets of E to subsets of $\mathbb{T}$, applying morphological operators on the lattice $\mathcal{P}(\mathbb{T})$, and then 'projecting' the results back to the original space E. The case of non-Boolean lattices, such as the lattice of gray value functions, requires the notion of sup-generating families (Heijmans & Ronse, 1989; Ronse & Heijmans, 1991; Heijmans, 1994). It turns out that only part of the invariance results for Boolean lattices carries over to the case of non-Boolean lattices with a non-abelian symmetry group; for details, see Roerdink (2000).

2.3 (Hyper)connected Filters

Connected filters are used to perform filtering based on various shape and size attributes (Salembier, Oliveras, & Garrido, 1998; Serra & Salembier, 1993; Salembier & Serra, 1995). A key property of connected filters is their edge preserving nature. Connected filters rely on an axiomatic definition of connectivity within a complete lattice framework (Serra, 1988; Heijmans, 1994, 1999; Braga-Neto & Goutsias, 2003). To address some shortcomings, hyperconnectivity and hyperconnected filters were introduced by Serra (1998). In particular, hyperconnected filters can deal with overlapping objects as separate entities (Serra, 1998; Wilkinson, 2007, 2009;

Ouzounis & Wilkinson, 2007, 2011) and can prevent the so-called "leakage" problem of connected filters. (Hyper)connected operators will be very useful for systematically simplifying complex oriented structures. Furthermore, connected filters are readily extended to the vector case (Salembier & Garrido, 2000), so it is expected that they extend to the tensor case as well.

2.4 Inf-Semilattices

We briefly mention Inf-semilattices here, as these are related to frames and sponges which will be discussed later.

The extension of mathematical morphology to complete semilattices was developed by Kresch (later called Keshet) (Kresch, 1998, 2000). The main motivation for this extension is that there are cases (such as the difference between two real functions) where the existence of both a least and a greatest element is not obvious. Then the notion of complete *semilattice* would be useful.

An *inf-semilattice* is a partially ordered set $\mathcal{P}$ where every two-element subset $\{a, b\}$ of $\mathcal{P}$ has an infimum $a \wedge b$, but not necessarily a supremum.[1] An inf-semilattice is *complete* when every non-empty subset $B \subseteq \mathcal{P}$ has an infimum $\bigwedge B$.

Erosions on complete inf-semilattices can be defined as operators that distribute over infima. The difference with erosions on complete lattices is that the condition of preservation of the greatest element is dropped. Erosions on complete inf-semilattices retain many properties of their counterparts on complete lattices: they are increasing, and an infimum of erosions is itself an erosion. Algebraic openings can be defined in the same way as for complete lattices. Also the notion of adjoint dilation in an inf-semilattice can be defined, although in a less straightforward way. See Kresch (2000) for further details.

3. FRAMES

Marginal processing of vector-valued images assumes that the image is decomposed into channels correspond to a certain basis, and processes the channels independently. This does not make sense for proper vectors, as the results will be biased by the choice of basis, and when considering something as a vector, it is usually important to have everything invariant

[1] A *sup-semilattice* is defined analogously.

to the choice of basis. However, instead of using a basis (a minimal set of vectors spanning the vector space), we can also pick a *frame*: a not necessarily minimal set of vectors spanning the vector space. A basis is thus a frame, but a frame might contain many more vectors, and can even be infinite for finite dimensional vector spaces. In particular, we can create a frame that contains all unit vectors, and thus essentially encompasses all possible bases (morphological operators are often invariant to scaling, so the magnitude of the vectors is usually irrelevant).

Another way of looking at this, is by considering the set of transformations of the basis or frame that should not matter. For example, the frame consisting of all unit vectors can also be viewed as the frame that is invariant to all rotations. Now we can see that, sometimes, instead of being invariant to all rotations, we may wish to be invariant only to certain rotations. For example, in an RGB color space, while within a plane of constant lightness it might be natural to pick any two vectors (corresponding to two hues) to span the space, it may not always make sense to be invariant to rotations that completely invert the order on luminance, for example. It is also possible to consider transformations other than rotations, but then we have to be extra careful about how we weigh the different vectors; an example is shown in Section 3.6.2.

3.1 Definitions

A real Hilbert space is a real vector space with a symmetric, bilinear, and positive definite binary operator '$\cdot$': the inner product. The inner product gives rise to a metric by putting $d(a, b) = \sqrt{(a - b) \cdot (a - b)}$, and a Hilbert space is complete with respect to this metric.

A frame (Christensen, 2008) is a set of vectors $\{f_i\}_{i \in \mathcal{I}}$ in a Hilbert space such that there are finite and positive constants A and B satisfying (for all a in the Hilbert space)

$$A \, \|a\|^2 \leq \|F \, a\|^2 \leq B \, \|a\|^2,$$

where $(F \, a)_i = f_i \cdot a$, and it is assumed that the range of F also admits an inner product, which is in practice not difficult to ensure. It should be noted that a frame has at least one dual frame $\{\hat{f}_i\}_{i \in \mathcal{I}}$, so that $\hat{F}^*$ is a left-inverse of F: $\hat{F}^* \circ F = \mathrm{id}$. One particularly important choice is the so-called canonical dual frame, for which $\hat{F}^*$ is the Moore-Penrose pseudoinverse of F (Ben-Israel & Greville, 1974).

3.2 Lattices and Group Invariance

First, let us motivate why a frame is needed in the first place. If we have a Hilbert space that is also a lattice, such that the lattice is invariant to addition and multiplication by positive scalars, then it can be shown that it essentially has to be based on having lattices on the individual coefficients in some basis, and then combining those lattices using a "direct or lexicographical product"; loosely translated, that means we are limited to product orders and lexicographical orders (or, possibly, combinations of the two). Given that lexicographical orders lead to discontinuous joins and meets (Chevallier & Angulo, 2014), we focus on product orders. Typically, the group of transformations to which a basis is invariant are fairly limited, and in particular: a basis is never invariant to the group of all rotations. Using a frame allows us to sidestep this problem, as we can construct (infinite) frames in which the transformations are represented by permutations (van de Gronde & Roerdink, 2013a, 2013b, 2014b).

Now, we could also look explicitly for Hilbert spaces that are lattices, such that the lattice is rotation invariant, rather than invariant to vector addition and multiplication by positive scalars. However, in two dimensions it is clear that these two things are related to each other through the mapping $(x, y) \mapsto (\cos(x)\, e^y, \sin(x)\, e^y)$ and its set-valued inverse, and in higher dimensions it is easy to see that things do not get any easier.

3.3 Constructing a Group Invariant Lattice

To construct a group invariant lattice based on a particular Hilbert space, simply pick a set of vectors $\{f_i\}_{i\in\mathcal{I}}$ that is invariant to the transformation group $\mathbb{T}$, verify that the chosen set of vectors is, in fact, a frame, and turn it into a product lattice, using the same lattice on all of the individual coefficients. Theorem 1 shows that this gives a lattice that is invariant to a representation of $\mathbb{T}$, while Proposition 1 shows that the lattice constructed on $\mathbb{R}^{\mathcal{I}}$ may not give rise to a lattice on the original Hilbert space, but still induces a partial order invariant to the transformation group.

Theorem 1. *If $\mathcal{H}$ is a Hilbert space, $\{f_i\}_{i\in\mathcal{I}}$ is a frame invariant to the transformation group $\mathbb{T}$ on $\mathcal{H}$, then any lattice on $\mathbb{R}$ gives rise to a lattice on $\mathbb{R}^{\mathcal{I}}$ that is invariant to a representation of $\mathbb{T}$ on $\mathbb{R}^{\mathcal{I}}$.*

Proof. If the frame is invariant to $\mathbb{T}$, then (due to each transformation being a bijection), $\mathbb{T}$ must act as a permutation on the frame. That is, for every $\tau \in \mathbb{T}$ there is a bijection $p_\tau : \mathcal{I} \to \mathcal{I}$ such that

$$\forall i \in \mathcal{I} : \tau(f_i) = f_{p_\tau(i)}.$$

This permutation can be used as a representation of $\mathbb{T}$ on $\mathbb{R}^{\mathcal{I}}$:

$$\forall i \in \mathcal{I} : f_i \cdot \tau(a) = f_{p_\tau(i)} \cdot a.$$

Clearly, if we construct a lattice on $\mathbb{R}^{\mathcal{I}}$ as a direct product of some lattice on $\mathbb{R}$, then this larger lattice is invariant to this representation of $\mathbb{T}$. □

Proposition 1. *If $\mathcal{H}$ is a Hilbert space, $\{f_i\}_{i\in\mathcal{I}}$ is a frame with the analysis operator F, and $\mathbb{R}^{\mathcal{I}}$ is a partially ordered set invariant to a representation of the transformation group $\mathbb{T}$ on $\mathcal{H}$, then this leads to a partial order on* **H** *invariant to $\mathbb{T}$ as follows: $a \le b \iff Fa \le Fb$ (for all a and b in $\mathcal{H}$).*

Proof. The given relation on $\mathcal{H}$ can be seen, from the definition, to inherit reflexivity, antisymmetry, and transitivity from the partial order on $\mathbb{R}^{\mathcal{I}}$, and is thus a partial order. Furthermore, consider ρ_τ to be the representation of $\tau \in \mathbb{T}$, and note that the partial order on $\mathbb{R}^{\mathcal{I}}$ is invariant to ρ_τ for all $\tau \in \mathbb{T}$. Now, for all $\tau \in \mathbb{T}$,

$$\begin{aligned} a \le b \iff Fa \le Fb \iff \rho_\tau(Fa) \le \rho_\tau(Fb) \iff F\tau(a) \le F\tau(b) \\ \iff \tau(a) \le \tau(b). \end{aligned}$$

This shows that the given partial order on $\mathcal{H}$ is invariant to the group $\mathbb{T}$. □

Example 1 (Rotation around axis)**.** Take the Euclidean space $\mathbb{R}^3$ as Hilbert space, and take as frame the set $\{\frac{1}{\sqrt{2}}(\cos(\alpha), \sin(\alpha), 1) \mid \alpha \in [0, 2\pi)\}$. Clearly, this frame is invariant to the group of rotations around the z-axis: $\{(0, 0, z) \mid z \in \mathbb{R}\}$. Theorem 1 now implies that any lattice on $\mathbb{R}$ gives rise to a lattice on $\mathbb{R}^{[0,2\pi)}$ that is invariant to a representation of the group of rotations around the z-axis. In particular, the lattice will be invariant to the representation that converts a rotation of β around the z-axis into a permutation that maps index α to $\alpha \pm \beta \mod 2\pi$, with the sign determined by the "handedness" of the rotation. Due to Proposition 1, if we pick the usual lattice on $\mathbb{R}$, the resulting lattice on $\mathbb{R}^{[0,2\pi)}$ induces a partial order on the original Hilbert space $\mathbb{R}^3$ that can be checked to be isomorphic to the Loewner order on 2×2 symmetric (real) matrices (van de Gronde & Roerdink, 2014a, §3.2).

It should be noted that instead of a lattice on the individual coefficients, we can also use a semilattice, like the one proposed by Heijmans and Keshet (2002), in which numbers are ordered by magnitude, with zero as the lowest number, and negative and positive numbers being incomparable. This approach is illustrated in the next example.

Example 2 (Complete rotation invariance). Take the Euclidean space $\mathbb{R}^3$ as Hilbert space, and take as frame the set $S_2 = \{v \mid \|v\| = 1\}$. Clearly, this frame is invariant to the group of all rotations $SO(3)$. Theorem 1 now implies that any lattice on $\mathbb{R}$ gives rise to a lattice on $\mathbb{R}^{S_2}$ that is invariant to a representation of $SO(3)$. In particular, the lattice will be invariant to the representation that converts a rotation $\tau \in SO(3)$ into a permutation that maps index v to $\tau^{-1}(v)$. If we pick the usual lattice on $\mathbb{R}$, the resulting lattice on $\mathbb{R}^{[0,2\pi)}$ induces a partial order on the original Hilbert space $\mathbb{R}^3$ that can be checked to be trivial, in the sense that no elements are comparable. In contrast, if we pick the inf-semilattice on $\mathbb{R}$ proposed by Heijmans and Keshet (2002), in which numbers are ordered by magnitude, with zero as the lowest number, and negative and positive numbers being incomparable, then the result is related to the inner product sponge discussed in Section 4.3.1 (also see the next section), although the partial order induced on $\mathbb{R}^3$ is still only mildly non-trivial: elements are only comparable if the lie along the same ray beginning at the origin.

3.4 Going Back

The frames resulting from the procedure explained in the previous section tend to be infinite, as most of the interesting groups tend to be infinite. And even if we stick to a finite sampling of such a frame, the number of vectors needed to give reasonable results will typically still be quite a bit higher than the number of vectors in a basis. As a result, it is typically desirable to somehow go back to a representation using a basis, and several options are possible.

Perhaps the most generic, and often easiest, option is to use a dual frame. In particular, if we use the Moore-Penrose pseudoinverse (corresponding to the canonical dual frame), we get a least squares solution, in the sense that $F^{\dagger}u = \arg\min_a \|u - Fa\|^2$. However, other dual frames could also be used, if this suits the application. To make this work we need an inner product on the space of frame coefficients, and it has to be invariant to the representation of the transformation group to which the lattice on the frame coefficients is invariant (van de Gronde & Roerdink, 2013b, §IV, 2014b). In other words, there must be a representation that maps the given transformation to the intersection of automorphisms (of the lattice on the frame coefficients) and orthogonal operations (with respect to the inner product on the frame coefficients).

Another possibility is to go back in such a way that the result is either an upper bound or a lower bound. To this end, imagine we have a function

$h : \mathcal{H} \to \mathbb{R}$ such that $Fa < Fb \implies h(a) < h(b)$, then we can define

$$P_+(u) = \arg\min_a h(a) \text{ subject to } u \leq Fa, \text{ and}$$
$$P_-(u) = \arg\max_a h(a) \text{ subject to } Fa \leq u.$$

It is often possible to give a suitable h that makes the above well-defined (such that there is always a solution, and it is unique). It can be verified that if the latter is the case, and

$$\forall \tau \in \mathbb{T} : h(a) \leq h(b) \implies h(\tau(a)) \leq h(\tau(b))$$

holds for the same transformation group $\mathbb{T}$ as the frame is invariant to, then $\tau \circ P_+ = P_+ \circ \rho_\tau$ for all $\tau \in \mathbb{T}$ (and similarly for P_-), and the representation (ρ) used in Theorem 1.

Example 3 (The Loewner pseudo-lattice)**.** Continuing from Example 1, computing $P_+(Fa \vee Fb)$, with $h(x, y, z) = z$, gives the pseudo-join based on the Loewner order, as used by Burgeth, Bruhn, Didas, Weickert, and Welk (2007). Similarly, $P_-(Fa \wedge Fb)$ gives the pseudo-meet. These operators are not proper joins and meets because they are not associative (van de Gronde & Roerdink, 2014a).

Example 4 (The inner product sponge)**.** Continuing from Example 2, computing $P_+(Fa \wedge Fb)$, with $h(x, y, z) = x^2 + y^2 + z^2$, gives the meet of the inner product sponge described in Section 4.3.1. Given that here the underlying structure is an inf-semilattice, we cannot really define the join in a similar way.

3.5 Practical Considerations

Many interesting groups are infinite, and this presents some issues when implementing the above. In particular, we obviously cannot directly compute an infinite number of inner products when computing Fa, nor can we simply store a list containing an infinite number of coefficients. There are at least two ways to get around this:

- approximate an infinite frame by a finite one, for example selecting just a few dozen uniformly distributed vectors can give very good results in practice;
- represent the frame coefficients implicitly, for example by using convex hull-related structures (van de Gronde, 2015, §5.3).

Figure 1 From left to right and top to bottom: the original, and the result of an erosion, dilation, and opening using a hue-invariant frame based on the usual lattice on the reals, with least squares backprojection. The images are processed at a resolution of 384×256 (half the original resolution), and a 5×5 square structuring element is used. (The parrot image is based on a photograph by Luc Viatour/www.Lucnix.be, used under the CC BY 2.0 license.)

The first option is easy, and typically gives good enough results, so is the one that was used so far in implementations. The second option is primarily interesting from a theoretical perspective, since it links frames to several other methods, but if high quality results are needed, it could also be faster/less memory intensive.

3.6 Examples

3.6.1 Hue or Rotation Invariant

Given an RGB color space, we can consider the group of hue rotations, which for the purposes of this example will be considered as the group of rotations around the gray axis. The construction of the frame follows Example 1. Fig. 1 shows the result on an image. Alternatively, we can consider all rotations in the RGB space. The construction of the frame follows Example 2. Fig. 2 shows the result on an image.

Figure 2 Top row: the result of an erosion and opening using a rotation-invariant frame based on an inf-semilattice, with least squares backprojection. Bottom row: the result of a dilation using least squares backprojection and least squares backprojection on valid values only. In the dilation result on the left there are many areas where the join is ill-defined for some coefficients, indicated by the bright and contrasting colors. On the right, this is avoided by only using valid coefficients (so for which the join did exist) in the least squares procedure; the result is qualitatively comparable to the dilation in Fig. 5.

3.6.2 Hue and Saturation Invariant

Changing the saturation of a color can be modeled in the RGB space as scaling that part of the color that is orthogonal to the gray axis. Such scalings form a group, and making use of the fact that (flat) morphological operators are themselves invariant to monotone transformations of individual coefficients in product lattices, it is possible to define a frame that allows defining operators that are invariant to both hue rotations and saturation changes (van de Gronde & Roerdink, 2013b). Fig. 3 illustrates the effect on an image.

4. SPONGES

Frames can successfully enable the application of traditional morphological concepts on vector- and tensor-valued data, but at the cost of

Figure 3 Top row: the result of an erosion and opening using a hue-and-saturation-invariant frame based on an inf-semilattice, with least squares backprojection. Bottom row: the result of a dilation using least squares backprojection and least squares backprojection on valid values only. In the dilation result on the left there are many areas where the join is ill-defined for some coefficients, indicated by the bright and contrasting colors. On the right, this is avoided by only using valid coefficients (so for which the join did exist) in the least squares procedure; the result is qualitatively comparable to the dilation in Fig. 4.

requiring a more complex representation of (intermediate) results. Apart from implying a longer processing time, this also means that when using frames we either have to save results in a different format then the input, or accept a certain amount of information loss, which can cause problems when chaining operations for example. An alternative is to use *sponges* (van de Gronde & Roerdink, 2015, 2016): generalizing lattices, sponges provide more flexibility in their definition, while still allowing us to recover at least some of the familiar results from lattice-based mathematical morphology.

4.1 Definitions

An oriented set is a set with a reflexive and antisymmetric relation '$\preceq$' (an orientation), such that $a \preceq b$ is interpreted as a being "less than or equal to" b in some way, without this relation necessarily being transitive. We write $A \preceq B$ for subsets A and B of an orientation if and only if $\forall a \in A, b \in$

$B : a \preceq b$. A sponge is an oriented set in which there exists a supremum (infimum) for every nonempty and finite subset with at least one upper bound (lower bound). That is, if A is a finite and nonempty subset of a sponge, then there exists a supremum of A if and only if there is an element that bounds all elements of P from above. By definition, the supremum of A (if it exists) is the least upper bound, in the sense that it is less than all other upper bounds.

Suprema in a sponge S are given through the join function $J : \mathcal{P}(S) \to \mathcal{P}^1(S)$, and infima through the meet function $M : \mathcal{P}(S) \to \mathcal{P}^1(S)$, where $P^1(S)$ gives all subsets of S of size at most one (so the empty set and all singletons). If $J(A) = \emptyset$ for some subset A, then A has no supremum, and thus no upper bounds. The functions J and M satisfy the following properties (for all $b \in S$ and A a finite, nonempty subset of the sponge S):

absorption: $\forall a \in A : M(\{a\} \cup J(A)) = \{a\}$,

part preservation:

$$[\forall a \in A : M(\{a, b\}) = \{b\}] \implies M(A) \neq \emptyset \text{ and } M(M(A) \cup \{b\}) = \{b\},$$

and the same properties with the roles of J and M reversed. Absorption is a direct adaptation of the same concept in lattices. Part preservation was adapted from weakly associative lattices (or trellises) (Skala, 1971; Fried & Grätzer, 1973a, 1973b) and specifies that any lower (upper) bound of a set is also a lower (upper) bound of the infimum (supremum) of this set. In a lattice, part preservation is implied by associativity. Commutativity is implied in sponges, as they operate on sets. Idempotence follows from combining the two absorption laws.[2] Note that the existence of a J and M with these properties is also sufficient to define a sponge.

A *tournament* is a totally oriented (sub)set. That is, it is an orientation (or subset of an orientation) such that every pair of elements is comparable. A subset A of a sponge S is *tournament-sup complete* if and only if $J(T) \subseteq A$ for every non-empty tournament $T \subseteq A$. Note that this definition does not require the supremum of every tournament to exist ($J(T)$ to be non-empty), but if it does exist in the original sponge, it should be in A. We briefly summarize some results derived in earlier work.

Proposition 2 (Prop. 2 in van de Gronde & Roerdink, 2016). *An acyclic tournament is a chain [a totally ordered set].*

[2] In the original definition J and M were partial functions, and idempotence had to be required explicitly. Here we follow more recent conventions introduced in joint work with Wim Hesselink (van de Gronde & Hesselink, submitted for publication).

As a consequence, if an orientation is acyclic, any subtournament is also acyclic, and thus a chain.

Proposition 3 (Prop. 3 in van de Gronde & Roerdink, 2016). *A tournament never has more than one maximal (minimal) element, and if it has a maximal (minimal) element it is the supremum (infimum) of the tournament.*

Proposition 4 (Prop. 4 in van de Gronde & Roerdink, 2016 – Hausdorff's Maximal Principle for orientations). *Every totally oriented subset T of an orientation O is contained in a maximal (in the sense of set inclusion) totally oriented subset M.*

The latter assumes the principle holds for partial orders, or, equivalently, that the axiom of choice holds (Birkhoff, 1961, §VIII.14).

Proposition 5 (Prop. 5 in van de Gronde & Roerdink, 2016). *A maximal totally oriented subset of a tournament-sup complete subset of a sponge contains its supremum, if it exists (in the sponge).*

Proposition 6 (Prop. 6 in van de Gronde & Roerdink, 2016). *In a conditionally complete sponge, the set of lower bounds $L(a)$ is tournament-sup complete for [every] a in the sponge.*

Proposition 7 (Prop. 7 in van de Gronde & Roerdink, 2016). *The intersection of two tournament-sup complete subsets of a sponge is tournament-sup complete.*

A subset A of a sponge S is *nonredundant* if and only if

$$a \preceq b \implies a = b, \quad \forall a, b \in A.$$

The *reduction operator* $\psi_{\mathcal{N}} : \mathcal{P}(S) \to \mathcal{P}(S)$ is defined by

$$\psi_{\mathcal{N}}(A) = \{a \mid a \in A \text{ and } \nexists b \in A : a \prec b\}.$$

A subset A is said to refine a subset B ($A \sqsubseteq B$) if and only if every element in A is bounded from above by an element from B. The result of the reduction operator is nonredundant, but need not (in general) be refined by the original set (situations where this *is* true play an important role in Section 4.2).

A sponge S is called *dry* if it satisfies (for all a, b in S and *nonredundant*[3] $A \subseteq S$, with $A \neq \emptyset$ and $J(A) \neq \emptyset$)

$$\{a\} \preceq J(A) \text{ and } A \preceq \{b\} \implies a \preceq b \text{ or } \left[\exists c \in A : a \preceq c\right]$$

If a sponge is not dry, it is *wet*.

4.2 Openings

In a lattice, an operator is considered an opening if it is anti-extensive, idempotent, and increasing. Increasingness likely cannot generalize directly to sponges, since even the join and meet are not necessarily "increasing", in the sense that $a \preceq c$ and $b \preceq d$ need not imply $J(\{a, b\}) \preceq J(\{c, d\})$, for example. However, we will see that it is possible to get familiar-looking operators that are anti-extensive and idempotent. Also, it is possible to replace increasingness by a weaker property that together with anti-extensivity and idempotence still characterizes openings in lattices (van de Gronde & Roerdink, 2016). The idea is that such a weaker property might be easier to apply to sponge-based operators, although so far this has not yet been shown to be true.

We start with a slightly reworked version of a result shown earlier (van de Gronde & Roerdink, 2016, Thm. 7). For this, consider the operator $\gamma_I^{\mathcal{N}} : S \to S$ on a conditionally complete sponge S, defined by

$$\gamma_I^{\mathcal{N}}(a) \in J(\psi_{\mathcal{N}}(\{b \mid b \in I \text{ and } b \preceq a\})),$$

with I a subset of S that contains a lower bound for every element in S.[4] It is not difficult to verify that $\gamma_I^{\mathcal{N}}$ is anti-extensive, in contrast to similar operators that are based on the Loewner order (van de Gronde & Roerdink, 2014a), which is not a sponge.

Theorem 2. *In a dry conditionally complete sponge S, $\gamma_I^{\mathcal{N}}$ is idempotent, assuming $I \cap L(a) \sqsubseteq \psi_{\mathcal{N}}(I \cap L(a))$ for all $a \in S$.*

Proof. Assume $f \in S$ is given. Define $A = \psi_{\mathcal{N}}(I \cap L(f))$ and $B = I \cap L(\gamma_I^{\mathcal{N}}(f))$, so $\gamma_I^{\mathcal{N}}(f) \in J(A)$ and $\gamma_I^{\mathcal{N}}(\gamma_I^{\mathcal{N}}(f)) \in J(\psi_{\mathcal{N}}(B))$. Since B is the

[3] Originally, also redundant sets were considered, but this is not necessary for Theorem 2, and makes it more complicated to prove that a sponge is dry.

[4] Here we do not require I to be tournament-sup complete, as this property is actually not needed in any of the proofs. In a complete lattice, $I \cap L(a) \sqsubseteq \psi_{\mathcal{N}}(I \cap L(a))$ follows from I being tournament-sup complete.

subset of I that is bounded by the join of A, itself a subset of I, $A \subseteq B$. Also, from the definitions of A and B, we get $\{b\} \preceq J(A)$ and $A \preceq \{f\}$ for all $b \in B$. Given that $I \cap L(f) \sqsubseteq \psi_{\mathcal{N}}(I \cap L(f))$, and $I \cap L(f)$ is guaranteed to be non-empty, A is non-empty. Now, since S is dry (and A is nonempty and nonredundant and $J(A) \neq \emptyset$), we have $b \preceq f$ or $\exists a \in A : b \preceq a$ for all $b \in B$.

We now show that $\psi_{\mathcal{N}}(B) \subseteq A$. If $b \in B$ satisfies $b \preceq f$, then it is in $I \cap L(f)$, and if it is not in A, then it is also not in $\psi_{\mathcal{N}}(B)$, since it is bounded from above by some element in $A \subseteq B$. Alternatively, if $\exists a \in A : b \preceq a$, then either $b \in A$ or $b \notin \psi_{\mathcal{N}}(B)$, since it is bounded from above by some element in $A \subseteq B$. Summarizing, every $b \in B$ must either be in A or not in $\psi_{\mathcal{N}}(B)$: $\psi_{\mathcal{N}}(B) \subseteq A$.

On the other hand, since $A \subseteq B$, and $B \sqsubseteq \psi_{\mathcal{N}}(B)$, if $a \in A \setminus \psi_{\mathcal{N}}(B)$, then we have a contradiction, because there then cannot be any element in $\psi_{\mathcal{N}}(B)$ that bounds a from above (given that $\psi_{\mathcal{N}}(B)$ is a subset of A, and that A is nonredundant), even though a is in B. Clearly $\psi_{\mathcal{N}}(B) = A$. As a consequence, $J(A) = J(\psi_{\mathcal{N}}(B))$ and $\gamma_I^{\mathcal{N}}(f) = \gamma_I^{\mathcal{N}}(\gamma_I^{\mathcal{N}}(f))$. □

It was shown previously that structural "openings" formed by combining a structural "erosion" and a structural "dilation" can be viewed as an operator $\gamma_I^{\mathcal{N}}$ for use with Theorem 2, with

$$I = \{a X_y \mid y \in E \text{ and } a \in S\}.$$

Here it is assumed that a sponge $\mathbf{Fun}(E, S)$ is used, with S a conditionally complete sponge with a least element, E a vector space, and $X \subset E$ a structuring element; X_y is used to denote $\{x + y \mid x \in X\}$, and $a X_y$ should be interpreted as in Eq. (4).

We now proceed to show that (trivial) connected openings based on flat zones can be defined based on a set I that can be used in Theorem 2 as well. A flat zone is simply a (connected) region of an image that has the same value everywhere within the region, and it will be represented as follows (for every position x in the domain E, and a (value) sponge S with a least element O_S, and an element $a \in S$):

$$(aX)(x) = \begin{cases} a & x \in X, \text{ and} \\ O_S & \text{otherwise.} \end{cases} \tag{4}$$

A connected opening $\gamma_I^{\mathcal{N}}$ can now be defined by picking as I a set of flat zones that satisfy certain criteria. For example, all flat zones whose support has an area of at least 100 pixels. It is assumed that the support sets of such flat zones are a suitable subset of a connectivity class (Heijmans, 1999), and in particular that the set of "valid" support sets $\mathcal{D}$ is closed for (infinite) unions of families of support sets that have a nonempty intersection:

$$\{X_i\}_{i\in\mathcal{I}} \subseteq \mathcal{D} \text{ and } \bigcap_{i\in\mathcal{I}} X_i \implies \bigcup_{i\in\mathcal{I}} X_i \in \mathcal{D}.$$

Theorem 3. *If S is a conditionally complete sponge with a least element, and $\mathcal{D}$ is a set of nonempty subsets of E, closed for infinite unions of families of support sets that have a nonempty intersection, then $I = \{aX \mid a \in S$ and $X \in \mathcal{D}\}$ satisfies $I \cap L(f) \sqsubseteq \psi_{\mathcal{N}}(I \cap L(f))$ for all $f \in \mathrm{Fun}(E, S)$.*

Proof. Note that a flat zone is greater than or equal to another flat zone if both the support and the associated value are greater than or equal to those of the other. For each support set $X \in \mathcal{D}$, $mX \in I \cap L(f)$, with $m \in M(\{f(x) \mid x \in X\})$. Furthermore, if $bY \in I \cap L(f)$, and $Y \supseteq X$, then $b \preceq m$. If there is a family of support sets $\{Y_i\} \subseteq \mathcal{D}$ such that $Y_i \supseteq X$ and $bY_i \in I \cap L(f)$ for all i, then $b(\cup\{Y_i\}) \in I \cap L(f)$ as well. As a consequence, for each support set $X \in \mathcal{D}$ there is a maximal element in $I \cap L(f)$ whose value is equal to the meet of the values in f within the support X, and whose support is $Y \in \mathcal{D}$ for some $Y \supseteq X$. The statement is now proved by noting that every element in $I \cap L(f)$ is of the form aX, with X a member of $\mathcal{D}$, and $a \in S$; given that $aX \preceq f$, we have $\{a\} \preceq \{f(x) \mid x \in X\}$, and xA is thus bounded from above by the maximal element corresponding to X (which has value $M(\{f(x) \mid x \in X\})$). □

Note that Theorem 2 does not necessarily apply to a sponge of the form $\mathrm{Fun}(E, S)$, where S is dry, as the direct product of dry sponges need not be dry. So, although we know that there are both dry sponges (see Section 4.3.1), and appropriate sets suitable for use with Theorem 2, it is not yet clear whether they indeed lead to idempotent openings. However, in practice it looks like sponges of the form $\mathrm{Fun}(E, S)$, where S is dry do indeed allow idempotent (structural) openings, so it remains to be seen whether it is simply very difficult to construct counterexamples, or if Theorem 2 can be further generalized (or changed).

4.3 Examples

4.3.1 The Inner Product Sponge

One of the main examples of a sponge is the so-called *inner product sponge*. The idea is quite simple: in an inner product (or, preferably, Hilbert) space,

$$a \preceq b \iff a \cdot (b - a) \geq 0.$$

Figure 4 From left to right and top to bottom: the original, and the result of an erosion, dilation, and "opening" using the direct product of the inner product sponge on the chroma plane, and a traditional (totally ordered) lattice along the lightness axis. The images are processed at half the original resolution, and a 5×5 square structuring element is used. Note that the dilation shows many artifacts in regions where different hues meet, as then it can be that there is no well defined supremum.

It has been shown that this indeed gives rise to a sponge, and that this sponge is dry in two dimensions.[5] Fig. 4 shows an example of such a sponge. Fig. 5 shows an example of using a three dimensional inner product sponge, illustrating that in practice it can still come quite close to exhibiting idempotent "openings", even if it is not dry.

4.3.2 The (Hemi)spherical Sponges

It is also possible to define sensible sponges on both a (hyper)sphere and a (hyper)hemisphere (van de Gronde & Roerdink, 2016), with the latter being interesting for working with discrete probability/likelihood distributions expressed in barycentric coordinates and mapped to a triangle on a (hyper)sphere by squaring the coordinates (Facchi, Kulkarni, Man'ko, Marmo, Sudarshan, & Ventriglia, 2010; Franchi & Angulo, 2015; Gromov,

[5] The proof given does not apply to all infinite sets, but in practice we only need it for finite sets.

Figure 5 From left to right and top to bottom: the result of an erosion, dilation, "opening", and "opening" of the "opening" using the inner product sponge directly on the RGB space. Note that since here the input data is strictly within the positive cone, it can be seen that a supremum always exists, so the dilation is effectively artifact free. In addition, although strictly speaking the "opening" need not be idempotent, in practice the results are quite close.

2013). However, these sponges can be related directly to the inner product sponge through various projections, so will not be discussed further here.

4.3.3 Periodic Sponges

A lattice is a partially ordered set, and given that a partial order is transitive, it cannot contain cycles. Sponges, however, can contain cycles. A simple example is $S = \mathbb{R}/\mathbb{Z}$, with $a \preceq b \iff \exists c \in a, d \in b : c \leq d < c + \frac{1}{2}$ for all $a, b \in S$. It can be verified that this relation is a dry sponge (van de Gronde & Roerdink, 2016, §5.5). In higher dimensions, the powers of S are also sponges, but not necessarily dry. In particular, if take S^3, then we see that if we take $a = [(0.6, 0.9, 1.0)]$, $P = \{[(0.8, 0.8, 1.0)], [(0.8, 1.0, 0.8)], [(1.0, 0.8, 0.8)]\}$, and $b = [(1.2, 1.2, 1.2)]$, that $a \preceq J(P)$ and $P \preceq b$, but also $a \npreceq b$, and $\forall p \in P : a \npreceq p$, the sponge is thus wet; here $[(a, b, c)] \in S^3$ can be identified with the set $\{(a + i, b + j, c + k) \mid i, j, k \in \mathbb{Z}\}$.

4.3.4 The Hyperbolic Sponge

One more sponge that should be mentioned, is the hyperbolic sponge (van de Gronde & Roerdink, 2016; van de Gronde & Hesselink, submitted for publication). This is a sponge on hyperbolic space (in any number of dimensions), in which sets of lower/upper bounds (depending on the point of view) are defined using geodesics, similar in spirit to sets of upper bounds in the inner product sponge (although it looks like the two are, in principle, not directly related). For examples of its application, see the work by Angulo and Velasco-Forero (2013, 2014) (which uses the same structure under a different name, in part because this work predates the identification of sponges).

5. *n*-ARY MORPHOLOGY

Another class of data for which it is difficult to pick a sensible total order is categorical data. After all, each category, or class, is simply different from all others. For this situation *n*-ary morphology was proposed: in this framework we no longer just have up and down, we have as many directions as we have categories. These categories could come from a per-pixel classification or an image segmentation, for example.

5.1 Definitions

In discrete *n*-ary morphology, an image is given as a function $f : \omega \to L$, where L is a set of labels, and Ω is some metric space with an (arbitrary) origin, and which allows translation; B_x will denote the translated version of B, such that if B contained the origin, B_x contains x. It will be assumed that B always contains the origin. The labels are assumed to be completely independent of each other. For each label $\ell \in L$, both a dilation δ_ℓ and an erosion ε_ℓ are defined. In the binary case, the dilation of one is the erosion of the other, and vice versa. The definition of the structural dilation with (flat) structuring element B presents few difficulties[6]:

$$\delta_B^\ell(f)(x) = \begin{cases} f(x) & \text{if } \forall y : x \in B_y \longrightarrow f(y) \neq \ell \\ \ell & \text{if } \exists y : x \in B_y \text{ and } f(y) = \ell. \end{cases}$$

[6] Compared to the original work, the reversed structuring element is applied. This is done to ensure that δ_B^ℓ and ε_B^ℓ form an adjunction in the preorder associated with ℓ.

The erosion is defined similarly, but requires a method for dealing with ambiguities, represented by the function θ_ℓ[7]:

$$\varepsilon_B^\ell(f)(x) = \begin{cases} f(x) & \text{if } f(x) \neq \ell \\ m & \text{if } \forall y \in B_x : f(y) = \ell \text{ or } f(y) = m \\ \theta_\ell(f, x, B) & \text{otherwise.} \end{cases}$$

Here, m stands for some arbitrary label satisfying the given condition, if one exists. At each position, at most one such m can exist. We put the following constraints on θ_ℓ:

1. θ_ℓ should always give a label other than ℓ;
2. $\theta_\ell(f, x, B)$ only depends on the values of f at the positions B_x (and their locations relative to x);
3. if $\theta_\ell(f, x, B) = m$, and g is an image such that $\forall x \in \Omega : f(x) = g(x)$ or $g(x) = m$, then $\theta_\ell(g, x, B) = m$ as well.

The effect of the above definition of the erosion is that

- values unequal to ℓ are not touched, since they are already "as far from ℓ as possible",
- if the set $f(B_x) = \{f(y) \mid y \in B_x\}$ only contains one label (which could be ℓ), then $\varepsilon_\ell(f, B)(x)$ will equal this label,
- if $f(B_x)$ contains ℓ and exactly one other label, then $\varepsilon_\ell(f, B)(x)$ will be equal to this other label,
- otherwise, we do not a priori know which label to pick, and θ will be used to make a choice.

Now, in the common case that B_x is a ball of a certain radius around x, the easiest (and, arguably, most correct (Chevallier, Chevallier, & Angulo, 2016, §3)) approach for defining θ_ℓ is to simply take the label assigned to the closest point to x that is not equal to ℓ. If this is not uniquely defined, we can either define some kind of arbitration rule, or assign a set of labels to the point, rather than a single label. Taking the closest label is exactly equivalent to considering the limit of implementing the erosion as repeated application of an erosion with a ball of smaller radius.[8]

[7] The second case is somewhat more general than in the original work: the original work only specified that the output should equal ℓ if $\forall y \in B_x : f(y) = \ell$. However, the more general rule provides a reasonable constraint on the erosion, and fits in well with the proposed choices for θ_ℓ.

[8] Strictly speaking this limit process may not always work, for example when applying the Euclidean metric to a discrete grid, but the idea still serves as motivation for defining θ to take the closest label.

5.2 Adjunctions, Openings, and Closings

Traditionally, each dilation has a corresponding erosion, such that the pair satisfies an adjunction:

$$\delta(f) \leq g \iff f \leq \varepsilon(g).$$

This is a very powerful relationship, that allows one to show, for example, that the composition of an erosion and a dilation is an opening or closing (depending on the order of the composition).

For n-ary morphology, instead of having a single partial order or lattice, we define a preorder $\leq_\ell$ consistent with δ_B^ℓ and ε_B^ℓ:

$$a \leq_\ell b \iff a \neq \ell \text{ or } b = \ell.$$

The consistency is captured in Lemma 1. Since a preorder is not necessarily anti-symmetric, it is possible that $a \leq_\ell b$ and $b \leq_\ell a$, but $a \neq b$, we say that the two are not equal but equivalent: $a \equiv_\ell b$. Since a preorder is transitive, the equivalence relation '$\sim_\ell$' is transitive as well, and we can recognize equivalence classes: subsets of L that are all equivalent to each other. In the current case there are two such classes: $\{\ell\}$ and $L \setminus \{\ell\}$.

Lemma 1. *If $y \in B_x$, then $f(x) \leq_\ell \delta_B^\ell(f)(y)$ and $\varepsilon_B^\ell(f)(x) \leq_\ell f(y)$.*

Proof. This follows directly from the definitions, the assumption made at the start of the section that B contains the origin, and the constraint that the case using θ should never give ℓ. □

Giving this consistency, we can show that a structural dilation and its corresponding erosion in the n-ary framework indeed form an adjunction.

Theorem 4. *δ_B^ℓ and ε_B^ℓ form an adjunction in the preorder $\leq_\ell$:*

$$\delta_B^\ell(f) \leq_\ell g \iff f \leq_\ell \varepsilon_B^\ell(g). \tag{5}$$

Proof. First, assume that $\delta_B^\ell(f) \leq_\ell g$ holds. Then, $f(x) \leq_\ell g(y)$ for all x and y with $y \in B_x$. As a result, for a given x, if $f(x) = \ell$, then $g(y) = \ell$ for all $y \in B_x$, and $\varepsilon_B^\ell(g)(x) = \ell$ as well, and $f(x) \leq \varepsilon_B^\ell(g)(x)$. On the other hand, if $f(x) \neq \ell$, then $f(x) \leq \varepsilon_B^\ell(g)(x)$ holds by definition. Since this argument holds for all x, $f \leq_\ell \varepsilon_B^\ell(g)$.

Now assume that $f \leq_\ell \varepsilon_B^\ell(g)$ holds. Then, $f(x) \leq_\ell g(y)$ for all x and y with $y \in B_x$. As a result, for a given y, if there exists an x such that $y \in B_x$ and $f(x) = \ell$, then $g(y) = \delta_B^\ell(f)(y) = \ell$. On the other hand, if (for a given y)

no such x can be found, $\delta^{\ell}_{B}(f)(y) = f(y)$, which is different from ℓ due to the assumption that B contains the origin. In either case, $\delta^{\ell}_{B}(f)(y) \leq g(y)$. Since this holds for all y, $\delta^{\ell}_{B}(f) \leq_{\ell} g$. This concludes the proof. □

Heijmans (1994, Prop. 3.14) proves several statements concerning the composition of a dilation and erosion that form an adjunction, Theorem 5 does much the same thing for the current setting.

Lemma 2. *The operators δ^{ℓ}_{B} and ε^{ℓ}_{B} are increasing in the preorder $\leq_{\ell}$.*

Proof. First consider the dilation δ^{ℓ}_{B} and preorder $\leq_{\ell}$. Note that $f \leq_{\ell} g$ if and only if $L_{\ell}(f) \subseteq L_{\ell}(g)$, where $L_{\ell}(f) = \{x \in \Omega \mid f(x) = \ell\}$. Now, it can be seen that $L_{\ell}(\delta^{\ell}_{B}(f)) = \delta_{B}(L_{\ell}(f))$, where δ_{B} is an ordinary binary dilation with structuring element B. Since a normal dilation is increasing, $f \leq_{\ell} g \implies \delta^{\ell}_{B}(f) \leq_{\ell} \delta^{\ell}_{B}(g)$, proving that δ^{ℓ}_{B} is increasing. That ε^{ℓ}_{B} is increasing in $\leq_{\ell}$ can be shown the same way. □

Theorem 5. *The n-ary dilation and erosion δ^{ℓ}_{B} and ε^{ℓ}_{B} (below simply δ and ε) satisfy the following:*

$$f \leq_{\ell} \varepsilon(\delta(f)), \tag{6}$$

$$\delta(\varepsilon(f)) \leq_{\ell} f, \tag{7}$$

$$\delta(\varepsilon(\delta(f))) = \delta(f), \tag{8}$$

Proof. Eqs. (6) and (7) are shown by substituting $\delta(f)$ for g and $\varepsilon(g)$ for f, respectively, in Eq. (5). To show Eq. (8), we first show

$$\delta(\varepsilon(\delta(f))) \sim_{\ell} \delta(f), \tag{9}$$

From Eqs. (6) and (7), and Lemma 2, we see that $\delta(\varepsilon(\delta(f))) \geq_{\ell} \delta(f)$ and $\delta(\varepsilon(\delta(f))) \leq_{\ell} \delta(f)$, hence Eq. (9) holds. Now, for all $x \in \Omega$, if $\delta(f)(x) = \ell$, then by Eq. (9), so is $\delta(\varepsilon(\delta(f)))(x)$. Otherwise, $\varepsilon(\delta(f))(x) = \delta(f)(x) \neq \ell$, and $\varepsilon(\delta(f))(y) \neq \ell$ for all y such that $x \in B_{y}$ (this from the definition of ε and the constraint that θ_{ℓ} never gives ℓ). As a consequence $\delta(\varepsilon(\delta(f)))(x) = \varepsilon(\delta(f))(x) = \delta(f)(x)$. Eq. (8) follows. □

Recall that an (algebraic) opening is an operator that is increasing, idempotent, and anti-extensive, and that a closing is increasing, idempotent, and *extensive*.

Corollary 1. *The operator $\varepsilon^{\ell}_{B} \circ \delta^{\ell}_{B}$ is a closing, and the operator $\delta^{\ell}_{B} \circ \varepsilon^{\ell}_{B}$ is an opening.*

  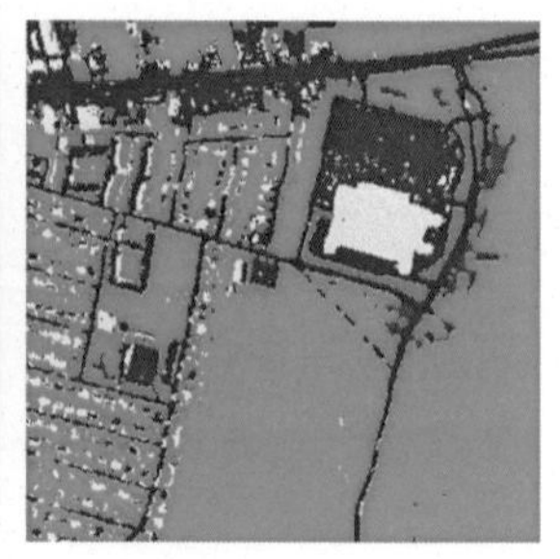

Figure 6 From left to right: the original, the dilation, and the erosion of the magenta label.

Figure 7 LEFT: opening of the magenta label. RIGHT: closing of the magenta label. See Fig. 6 for the original.

Proof. The composition of two increasing operators is increasing, hence both operators mentioned are increasing by Lemma 2. By Eqs. (6) and (7) of Theorem 5 the two operators are also extensive and anti-extensive, respectively. Finally, based on Eq. (8), it is quickly verified that both operators are idempotent, either by substituting $\varepsilon_B^\ell(f)$ for f, or by applying ε_B^ℓ to both sides of the equation. □

5.3 Example

This example uses the "Urban" hyperspectral data set (Zhu, Wang, Xiang, Fan, & Pan, 2014), where each pixel is labeled according to the largest abundance out of six classes at that point.[9] Fig. 6 shows the original, as well as the basic operations dilation and erosion. Fig. 7 shows the opening and closing of a particular label.

[9] Retrieved, including abundances, from http://www.escience.cn/people/feiyunZHU/Dataset_GT.html (2017-04-30).

6. OTHER APPROACHES

In addition to sponges and n-ary morphology, there have been more approaches to achieve qualitatively similar results to traditional morphology in nonscalar settings. For tensor-valued images, the Loewner order has been used by Burgeth, Bruhn, Didas, et al. (2007) to define erosion-, dilation-, opening-, and closing-like operators. The operators allow for relatively efficient implementations, but do violate some of the basic properties that define the traditional operators (van de Gronde & Roerdink, 2014a).

Another approach is to observe that the usual maximum on reals (at least on the positive reals) is intimately related to certain types of norms. Attempts have been made to do morphology based on generalizing such constructions to non-scalar settings (Burgeth, Welk, Feddern, & Weickert, 2004; Angulo, 2013). It is not yet clear if, and if so how, the results relate to other approaches, nor what properties the resulting operators preserve.

Related to rotation-invariant frame-based methods and inner product sponges, are methods based directly on convex hulls (Gimenez & Evans, 2005), as well as the vector median filter (Astola, Haavisto, & Neuvo, 1990).

Instead of considering pixel values, we can also associate values to the edges between pixels and use this to do morphology (especially connected morphology) (Chung & Sapiro, 2000; Salembier & Garrido, 2000; Aptoula & Lefèvre, 2007a; Rittner, Flores, & Lotufo, 2007; Rittner & Lotufo, 2008; Tarabalka, Chanussot, & Benediktsson, 2010; Alonso-González, Valero, Chanussot, López-Martínez, & Salembier, 2013; Valero, Salembier, & Chanussot, 2013). Alternatively, we can try impose orders that depend on the data (based on statistical principles for example) (Goutsias, Heijmans, & Sivakumar, 1995; Trahanias & Venetsanopoulos, 1996; Lezoray, Elmoataz, & Ta, 2008; Velasco-Forero & Angulo, 2011a, 2011b, 2013).

When the data at hand can be interpreted as a vector- or tensor-field, so when the value at each position is a vector or tensor that can usefully be interpreted in terms of the tangent space of the image domain, then it may make sense to interpret these vectors and tensors more as giving scalar values in certain directions rather than as opaque non-scalar objects. This leads to path-based morphology (van de Gronde, Lysenko, & Roerdink, 2015; van de Gronde, 2015).

Finally, it should be mentioned that marginal processing (for example using a product lattice on a basis for vector-valued data) is often quite adequate, especially if it does not really make sense to mix the different components (Serra, 1993). Alternatively, one can use Pareto morphology in such situations (Köppen & Franke, 2007; Köppen, Nowack, & Rösel, 1999), even though this will give less stable results than marginal processing.

7. SUMMARY AND CONCLUSIONS

In this paper we have discussed extensions of mathematical morphology beyond the classical cases of binary and grayscale images with the standard ordering of sets or real function values, for which the complete lattice framework is no longer adequate. In particular, for multidimensional images such as vector and tensor valued images, as well as categorical data there is no obvious ordering of the value domain.

Nevertheless, morphological operators on such images can be defined by focusing on the invariance properties one would like to maintain, such as invariance under rotations or scaling. Taking inspiration from the concepts defined in group morphology (Roerdink, 2000), we have shown that this can be achieved by representing vectors or tensors using a *frame* rather than a basis. For example, in the case of color images we saw that in order to avoid the false color problem one should not use a particular color basis, but simply consider all possible bases generated by the required invariance (for example all rotations of a specific RGB basis), perform some morphological operator in this lifted space, and in the end project the result back to the original domain. This more complicated representation comes at a price, leading to a possible loss of information, as well as a loss of the properties that usually characterize morphological operators. This motivated the introduction of *sponges*, which are easier to adapt to multidimensional spaces, and allow us to recover parts of the usual morphological theory based on lattices.

For categorical data, or data with a vector of likelihoods for each category, we have shown how the concept of n-ary morphology can be used. Finally, we discussed various other approaches that have ambitions similar to the methods in this work, but either sacrifice (even more) properties, or take a completely different point of view.

REFERENCES

Alonso-González, A., Valero, S., Chanussot, J., López-Martínez, C., & Salembier, P. (2013). Processing multidimensional SAR and hyperspectral images with binary partition tree. *Proceedings of the IEEE*, *101*(3), 723–747. http://dx.doi.org/10.1109/jproc.2012.2205209.

Angulo, J. (2007). Morphological colour operators in totally ordered lattices based on distances: Application to image filtering, enhancement and analysis. *Computer Vision and Image Understanding*, *107*(1–2), 56–73. http://dx.doi.org/10.1016/j.cviu.2006.11.008.

Angulo, J. (2013). Supremum/infimum and nonlinear averaging of positive definite symmetric matrices. In F. Nielsen, & R. Bhatia (Eds.), *Matrix information geometry* (pp. 3–33). Berlin, Heidelberg: Springer.

Angulo, J., & Velasco-Forero, S. (2013). Complete lattice structure of Poincaré upper-half plane and mathematical morphology for hyperbolic-valued images. In F. Nielsen, & F. Barbaresco (Eds.), *LNCS: Vol. 8085. Geometric science of information* (pp. 535–542). Berlin, Heidelberg: Springer.

Angulo, J., & Velasco-Forero, S. (2014). Morphological processing of univariate Gaussian distribution-valued images based on Poincaré upper-half plane representation. In F. Nielsen (Ed.), *Geometric theory of information, signals and communication technology* (pp. 331–366). Springer International Publishing.

Aptoula, E., & Lefèvre, S. (2007a). A basin morphology approach to colour image segmentation by region merging. In Y. Yagi, S. Kang, I. Kweon, & H. Zha (Eds.), *LNCS: Vol. 4843. Computer vision* (pp. 935–944). Berlin, Heidelberg: Springer, ISBN 978-3-540-76385-7. Chapter 89.

Aptoula, E., & Lefèvre, S. (2007b). A comparative study on multivariate mathematical morphology. *Pattern Recognition*, *40*(11), 2914–2929. http://dx.doi.org/10.1016/j.patcog.2007.02.004.

Astola, J., Haavisto, P., & Neuvo, Y. (1990). Vector median filters. *Proceedings of the IEEE*, *78*(4), 678–689. http://dx.doi.org/10.1109/5.54807.

Ben-Israel, A., & Greville, T. N. (1974). *Generalized inverses: Theory and applications*. New York: Wiley, ISBN 0471065773.

Birkhoff, G. (1961). *Lattice theory. American mathematical society colloquium publications: Vol. 25*. American Mathematical Society.

Braga-Neto, U., & Goutsias, J. (2003). A theoretical tour of connectivity in image processing and analysis. *Journal of Mathematical Imaging and Vision*, *19*, 5–31.

Burgeth, B., Bruhn, A., Didas, S., Weickert, J., & Welk, M. (2007). Morphology for matrix data: Ordering versus PDE-based approach. *Image and Vision Computing*, *25*(4), 496–511. http://dx.doi.org/10.1016/j.imavis.2006.06.002.

Burgeth, B., Bruhn, A., Papenberg, N., Welk, M., & Weickert, J. (2007). Mathematical morphology for matrix fields induced by the Loewner ordering in higher dimensions. *Signal Processing*, *87*(2), 277–290.

Burgeth, B., & Kleefeld, A. (2017). A unified approach to PDE-driven morphology for fields of orthogonal and generalized doubly-stochastic matrices. In *Mathematical morphology and its applications to signal and image processing* (pp. 284–295).

Burgeth, B., Papenberg, N., Bruhn, A., Welk, M., Feddern, C., & Weickert, J. (2005). Morphology for higher-dimensional tensor data via Loewner ordering. In C. Ronse, L. Najman, & E. Decencière (Eds.), *Computational imaging and vision: Vol. 30. Mathematical morphology: 40 years on (proceedings of the 7th international symposium on mathematical morphology)* (pp. 407–416). Wien, New York: Springer.

Burgeth, B., Welk, M., Feddern, C., & Weickert, J. (2004). Morphological operations on matrix-valued images. In T. Pajdla, & J. Matas (Eds.), *LNCS: Vol. 3024. Computer vision* (pp. 155–167). Berlin, Heidelberg: Springer.

Cheng, F., & Venetsanopoulos, A. N. (2000). Adaptive morphological operators, fast algorithms and their applications. *Pattern Recognition*, *33*, 917–933.

Chevallier, E., & Angulo, J. (2014). The discontinuity issue of total orders on metric spaces and its consequences for mathematical morphology. Technical report, Centre de Morphologie Mathématique, MINES ParisTech.

Chevallier, E., Chevallier, A., & Angulo, J. (2016). N-ary mathematical morphology. *Mathematical Morphology Theory and Applications*, *1*(1). http://dx.doi.org/10.1515/mathm-2016-0003.

Christensen, O. (2008). *Frames and bases: An introductory course*. Birkhäuser/Springer, ISBN 9780817646783.

Chung, D. H., & Sapiro, G. (2000). On the level lines and geometry of vector-valued images. *IEEE Signal Processing Letters*, 7(9), 241–243. http://dx.doi.org/10.1109/97.863143.

Ćurić, V., Landström, A., Thurley, M. J., & Hendriks, C. L. L. (2014). Adaptive mathematical morphology – A survey of the field. *Pattern Recognition Letters*, *47*(0), 18–28. http://dx.doi.org/10.1016/j.patrec.2014.02.022. Advances in Mathematical Morphology.

Facchi, P., Kulkarni, R., Man'ko, V. I., Marmo, G., Sudarshan, E. C. G., & Ventriglia, F. (2010). Classical and quantum Fisher information in the geometrical formulation of quantum mechanics. *Physics Letters A*, *374*(48), 4801–4803. http://dx.doi.org/10.1016/j.physleta.2010.10.005.

Franchi, G., & Angulo, J. (2015). Ordering on the probability simplex of endmembers for hyperspectral morphological image processing. In J. A. Benediktsson, J. Chanussot, L. Najman, & H. Talbot (Eds.), *LNCS: Vol. 9082. Mathematical morphology and its applications to signal and image processing* (pp. 410–421). Springer International Publishing.

Fried, E., & Grätzer, G. (1973a). A nonassociative extension of the class of distributive lattices. *Pacific Journal of Mathematics*, *49*(1), 59–78.

Fried, E., & Grätzer, G. (1973b). Some examples of weakly associative lattices. *Colloquium Mathematicum*, *27*, 215–221.

Giardina, C. R., & Dougherty, E. R. (1988). *Morphological methods in image and signal processing*. Englewood Cliffs, NJ: Prentice Hall.

Gierz, G., Hofmann, K. H., Keimel, K., Lawson, J. D., Mislove, M., & Scott, D. S. (1980). *A compendium of continuous lattices*. Wien, New York: Springer.

Gimenez, D., & Evans, A. N. (2005). Colour morphological scale-spaces from the positional colour sieve. In *Digital image computing: Techniques and applications* (p. 60). IEEE, ISBN 0-7695-2467-2.

Goutsias, J., Heijmans, H. J. A. M., & Sivakumar, K. (1995). Morphological operators for image sequences. *Computer Vision and Image Understanding*, *62*(3), 326–346. http://dx.doi.org/10.1006/cviu.1995.1058.

Gromov, M. (2013). In a Search for a Structure, Part 1: On Entropy. Retrieved on 2015-06-18.

van de Gronde, J. J. (2015). Beyond scalar morphology. PhD thesis, University of Groningen.

van de Gronde, J.J., & Hesselink, W.H. (2016). Conditionally complete sponges: New results on generalized lattices. Submitted for publication.

van de Gronde, J. J., Lysenko, M., & Roerdink, J. B. T. M. (2015). Path-based mathematical morphology on tensor fields. In I. Hotz, & T. Schultz (Eds.), *Mathematics and Visualization. Visualization and processing of higher order descriptors for multi-valued data* (pp. 109–127). Springer International Publishing.

van de Gronde, J. J., & Roerdink, J. B. T. M. (2013a). Frames for tensor field morphology. In F. Nielsen, & F. Barbaresco (Eds.), *LNCS: Vol. 8085. Geometric science of information* (pp. 527–534). Berlin, Heidelberg: Springer.

van de Gronde, J. J., & Roerdink, J. B. T. M. (2013b). Group-invariant frames for colour morphology. In C. L. Hendriks, G. Borgefors, & R. Strand (Eds.), *LNCS: Vol. 7883. Mathematical morphology and its applications to signal and image processing* (pp. 267–278). Berlin, Heidelberg: Springer.

van de Gronde, J. J., & Roerdink, J. B. T. M. (2014a). Frames, the Loewner order and eigendecomposition for morphological operators on tensor fields. *Pattern Recognition Letters, 47*, 40–49. http://dx.doi.org/10.1016/j.patrec.2014.03.013.

van de Gronde, J. J., & Roerdink, J. B. T. M. (2014b). Group-invariant colour morphology based on frames. *IEEE Transactions on Image Processing, 23*(3), 1276–1288. http://dx.doi.org/10.1109/tip.2014.2300816.

van de Gronde, J. J., & Roerdink, J. B. T. M. (2015). Sponges for generalized morphology. In J. A. Benediktsson, J. Chanussot, L. Najman, & H. Talbot (Eds.), *LNCS: Vol. 9082. Mathematical morphology and its applications to signal and image processing* (pp. 351–362). Springer International Publishing.

van de Gronde, J. J., & Roerdink, J. B. T. M. (2016). Generalized morphology using sponges. *Mathematical Morphology Theory and Applications, 1*(1). http://dx.doi.org/10.1515/mathm-2016-0002.

Heijmans, H. J. A. M. (1987). Mathematical morphology: An algebraic approach. *CWI Newsletter, 14*, 7–27.

Heijmans, H. J. A. M. (1994). *Morphological image operators. Advances in electronics and electron physics: Vol. 25.* Academic Press, ISBN 0120145995.

Heijmans, H. J. A. M. (1999). Connected morphological operators for binary images. *Computer Vision and Image Understanding, 73*(1), 99–120. http://dx.doi.org/10.1006/cviu.1998.0703.

Heijmans, H. J. A. M., & Keshet, R. (2002). Inf-semilattice approach to self-dual morphology. *Journal of Mathematical Imaging and Vision, 17*(1), 55–80. http://dx.doi.org/10.1023/a:1020726725590.

Heijmans, H. J. A. M., & Ronse, C. (1989). The algebraic basis of mathematical morphology. Part I: Dilations and erosions. *Computer Vision, Graphics, and Image Processing, 50*, 245–295.

Köppen, M., & Franke, K. (2007). Pareto-dominated hypervolume measure: An alternative approach to color morphology. In *International conference on hybrid intelligent systems* (pp. 234–239). IEEE, ISBN 978-0-7695-2946-2.

Köppen, M., Nowack, C., & Rösel, G. (1999). Pareto-morphology for color image processing. In *Scandinavian conference on image analysis, Vol. 1* (pp. 192–202). ISBN 87-88306-42-9.

Kresch, R. (1998). Extensions of morphological operations to complete semilattices and its applications to image and video processing. In H. J. A. M. Heijmans, & J. B. T. M. Roerdink (Eds.), *Mathematical morphology and its application to image and signal processing* (pp. 35–42). Dordrecht: Kluwer Academic Publishers.

Kresch, R. K. (2000). Mathematical morphology on complete semilattices and its applications to image processing. *Fundamenta Informaticae, 41*(1–2), 33–56.

Ledoux, A., & Richard, N. (2016). Color and multiscale texture features from vectorial mathematical morphology. *Signal, Image and Video Processing, 10*(3), 431–438. http://dx.doi.org/10.1007/s11760-015-0759-3.

Lerallut, R., Decencière, E., & Meyer, F. (2007). Image filtering using morphological amoebas. *Image and Vision Computing, 25*(4), 395–404.
Lezoray, O., Elmoataz, A., & Ta, V. T. (2008). Learning graph neighborhood topological order for image and manifold morphological processing. In *International conference on computer and information technology* (pp. 350–355). IEEE, ISBN 978-1-4244-2357-6.
Maragos, P., & Vachier, C. (2009). Overview of adaptive morphology: Trends and perspectives. In *Proceedings of the ICIP'2009 (international conference on image processing).*
Matheron, G. (1975). *Random sets and integral geometry*. New York, NY: John Wiley & Sons.
Ouzounis, G. K., & Wilkinson, M. H. F. (2007). Mask-based second generation connectivity and attribute filters. *IEEE Transactions on Pattern Analysis and Machine Intelligence, 29*, 990–1004.
Ouzounis, G. K., & Wilkinson, M. H. F. (2011). Hyperconnected attribute filters based on k-flat zones. *IEEE Transactions on Pattern Analysis and Machine Intelligence, 33*(2), 224–239. http://dx.doi.org/10.1109/TPAMI.2010.74.
Rittner, L., Flores, F., & Lotufo, R. (2007). New tensorial representation of color images: Tensorial morphological gradient applied to color image segmentation. In *Brazilian symposium on computer graphics and image processing* (pp. 45–52). IEEE, ISBN 978-0-7695-2996-7.
Rittner, L., & Lotufo, R. (2008). Diffusion tensor imaging segmentation by watershed transform on tensorial morphological gradient. In *Brazilian symposium on computer graphics and image processing* (pp. 196–203). IEEE, ISBN 978-0-7695-3358-2.
Robinson, D. J. S. (1982). *A course in the theory of groups*. Wien, New York: Springer.
Roerdink, J. B. T. M. (1993). Mathematical morphology with non-commutative symmetry groups. In E. R. Dougherty (Ed.), *Mathematical morphology in image processing* (pp. 205–254). New York, NY: Marcel Dekker. Chapter 7.
Roerdink, J. B. T. M. (2000). Group morphology. *Pattern Recognition, 33*(6), 877–895. http://dx.doi.org/10.1016/S0031-3203(99)00152-1.
Roerdink, J. B. T. M. (2009). Adaptivity and group invariance in mathematical morphology. In *Proceedings of the ICIP'2009 (international conference on image processing)* (pp. 2253–2256).
Roerdink, J. B. T. M., & Heijmans, H. J. A. M. (1988). Mathematical morphology for structures without translation symmetry. *Signal Processing, 15*, 271–277. http://dx.doi.org/10.1016/0165-1684(88)90017-5.
Ronse, C., & Heijmans, H. J. A. M. (1991). The algebraic basis of mathematical morphology. Part II: Openings and closings. *Computer Vision, Graphics, and Image Processing: Image Understanding, 54*, 74–97.
Salembier, P., & Garrido, L. (2000). Binary partition tree as an efficient representation for image processing, segmentation, and information retrieval. *IEEE Transactions on Image Processing, 9*(4), 561–576. http://dx.doi.org/10.1109/83.841934.
Salembier, P., Oliveras, A., & Garrido, L. (1998). Anti-extensive connected operators for image and sequence processing. *IEEE Transactions on Image Processing, 7*, 555–570.
Salembier, P., & Serra, J. (1995). Flat zones filtering, connected operators, and filters by reconstruction. *IEEE Transactions on Image Processing, 4*, 1153–1160.
Serra, J. (1982). *Image analysis and mathematical morphology*. New York: Academic Press.
Serra, J. (Ed.). (1988). *Image analysis and mathematical morphology. II: Theoretical advances*. New York: Academic Press.
Serra, J. (1993). Anamorphoses and function lattices. In E. R. Dougherty, P. D. Gader, & J. C. Serra (Eds.), *SPIE proceedings: Vol. 2030. Image algebra and morphological image processing IV* (pp. 2–11).

Serra, J. (1998). Connectivity on complete lattices. *Journal of Mathematical Imaging and Vision, 9*(3), 231–251.

Serra, J., & Salembier, P. (1993). Connected operators and pyramids. In *SPIE image algebra morphological image processing IV: Vol. 2030* (pp. 65–76).

Skala, H. L. (1971). Trellis theory. *Algebra Universalis, 1*(1), 218–233. http://dx.doi.org/10.1007/bf02944982.

Soille, P. (2003). *Morphological image analysis* (2nd edition). Heidelberg: Springer-Verlag, ISBN 3-540-42988-3.

Suzuki, M. (1982). *Group theory*. Wien, New York: Springer.

Tarabalka, Y., Chanussot, J., & Benediktsson, J. A. (2010). Segmentation and classification of hyperspectral images using watershed transformation. *Pattern Recognition, 43*(7), 2367–2379. http://dx.doi.org/10.1016/j.patcog.2010.01.016.

Trahanias, P. E., & Venetsanopoulos, A. N. (1996). Vector order statistics operators as color edge detectors. *IEEE Transactions on Systems, Man and Cybernetics, Part B (Cybernetics), 26*(1), 135–143. http://dx.doi.org/10.1109/3477.484445.

Valero, S., Salembier, P., & Chanussot, J. (2013). Hyperspectral image representation and processing with binary partition trees. *IEEE Transactions on Image Processing, 22*(4), 1430–1443. http://dx.doi.org/10.1109/tip.2012.2231687.

Velasco-Forero, S., & Angulo, J. (2011a). Mathematical morphology for vector images using statistical depth. In P. Soille, M. Pesaresi, & G. K. Ouzounis (Eds.), *LNCS: Vol. 6671. Mathematical morphology and its applications to image and signal processing* (pp. 355–366). Berlin, Heidelberg: Springer, ISBN 978-3-642-21568-1. Chapter 31.

Velasco-Forero, S., & Angulo, J. (2011b). Supervised ordering in $\mathbb{R}^p$: Application to morphological processing of hyperspectral images. *IEEE Transactions on Image Processing, 20*(11), 3301–3308. http://dx.doi.org/10.1109/tip.2011.2144611.

Velasco-Forero, S., & Angulo, J. (2013). Supervised morphology for structure tensor-valued images based on symmetric divergence kernels. In F. Nielsen, & F. Barbaresco (Eds.), *LNCS: Vol. 8085. Geometric science of information* (pp. 543–550). Berlin, Heidelberg: Springer.

Wilkinson, M. H. F. (2007). Attribute-space connectivity and connected filters. *Image and Vision Computing, 25*, 426–435.

Wilkinson, M. H. F. (2009). An axiomatic approach to hyperconnectivity. In M. H. F. Wilkinson, & J. B. T. M. Roerdink (Eds.), *LNCS: Vol. 5720. Proceedings of the ISMM 2009* (pp. 35–46).

Zhu, F., Wang, Y., Xiang, S., Fan, B., & Pan, C. (2014). Structured sparse method for hyperspectral unmixing. *ISPRS Journal of Photogrammetry and Remote Sensing, 88*, 101–118.

CHAPTER FOUR

Energy Analyzing and Energy Selecting Electron Microscopes*

Allen J.F. Metherell
Formerly Cavendish Laboratory, Cambridge, England
e-mail address: allenmetherell@bellsouth.net

Contents

1. INTRODUCTION

CONTRAST is observed in an electron microscope image as a result of the fact that electrons in the incident beam are scattered by the specimen under examination. The scattering processes can be conveniently divided into two classes; elastic scattering in which the incident electron suffers no energy change, and inelastic scattering in which energy is lost to the specimen. For crystalline materials there is only one mechanism responsible for elastic scattering and that is diffraction or Bragg scattering of the incident electrons. The types of inelastic scattering which can occur are, however, numerous. It is possible for the fast electrons to lose energy by the excitation of a crystal quasi-particle such as a phonon, an exciton, a plasmon, etc. or by atomic ionization, by the scattering of atomic electrons from the valence band into the conduction band, by the scattering of atomic

* Reprinted from Advances in Optical and Electron Microscopy volume 4 (1971) 263–360.

Advances in Imaging and Electron Physics, Volume 204
ISSN 1076-5670
https://doi.org/10.1016/bs.aiep.2017.09.003

electrons from a given state in the conduction band to an empty state in the same band, etc. All inelastic scattering processes excite the specimen from its ground state energy level into higher energy states, and in principle it is possible to study the complicated energy level structure of a solid by measuring the energy loss spectrum of the incident fast electrons. Suppose that a specimen is being examined whose composition varies with position across the field of view. The composition changes are expected to be accompanied by changes in the energy level structure of the solid and hence electrons scattered from different regions of the specimen will have loss spectra characteristic of those regions. Measurement of the loss spectra of electrons arriving at different points in an electron microscope image should therefore provide a powerful method of microanalysis.

The conventional electron micrograph does not distinguish between electrons of different energies, and is therefore the electron optical analogue of a black and white photograph. It is pertinent to consider the possibility of constructing the electron optical analogue of a color photograph, because a micrograph which is sensitive to the electron energy utilizes all the physical information contained in the final image. The energy range of the scattered electrons is obviously an important factor which must be considered. The use of a "stop" or objective aperture in an electron microscope restricts the angles of scatter of the inelastically scattered electrons, which contribute directly to the image intensity, to values of the order of a Bragg angle (typically $\sim 10^{-2}$ rad for 100 keV electrons). This restriction reduces the number of inelastic scattering processes which have to be considered, since the majority of the observed energy loss electrons will be produced by loss mechanisms which possess angular distributions sharply peaked in the direction of the beam chosen to form the image. Furthermore only those loss processes with mean free paths of the order of the specimen thickness need be considered. Mean free paths for atomic ionization for example are typically of the order of millimeters, whereas mean free paths for plasmon excitation are typically of the order of thousands of Ångstrom units. Although both processes produce angular distributions of the scattered electrons sharply peaked in the forward direction, the proportion of electrons which suffer energy losses through atomic ionization are negligible in comparison with those which lose energy by plasmon excitation. The most predominant energy loss processes which contribute directly to the inelastic scattering observed in an electron microscope image are (a) plasmon excitations, possibly modified strongly by interband transitions, (b) single electron interactions corresponding to interband transitions

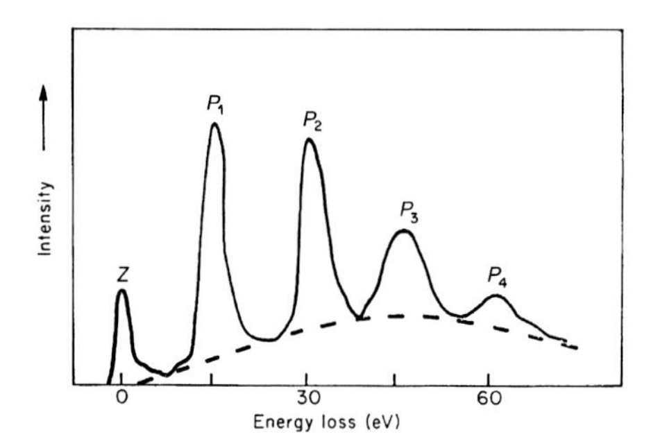

Figure 1 A typical example of the energy loss spectrum of electrons contributing to an electron microscope image of aluminum of thickness ~2000 Å and with an incident beam potential of 80 kV.

and (c) phonon excitations. A typical example of the loss spectrum of electrons contributing to the image of a specimen of aluminum is shown in Fig. 1. The broken line shows the loss profile due to single electron scattering, superimposed on which are the well defined plasmon losses P_1, P_2, etc. Clearly most of the energy losses suffered by the incident electrons lie within several hundred electron volts of the incident beam potential, which is typically of the order of hundreds of kilovolts, and the fractional change in the electron energy is therefore very small.

The difficulties involved in constructing the electron optical analogue of a color photograph stem mainly from the relatively small differences in the energies of the electrons forming the image of the specimen. The idealized energy analyzing electron microscope would record simultaneously the loss spectra from all points x, y (Fig. 2) lying within the field of view. This could only be achieved by displacing electrons of different energies in the z direction by amounts proportional to their energy losses. This within itself represents no real instrumental difficulty. The problem arises in finding a recording device which can measure the intensity variation simultaneously at all points within the volume xyz.

The lack of a suitable recording device of this type has led to the development of two instruments, the energy analyzing electron microscope and the energy selecting electron microscope, which utilize in different ways the information contained in the loss spectra of the electrons forming an electron microscope image. In the energy analyzing electron microscope a fine slit is placed in the final image plane of the microscope, so that in effect a line is selected in the image for energy analysis. The slit is made the entrance aperture of an electron spectrometer and electrons of various energies arriving at the image along this line are displaced in a direction

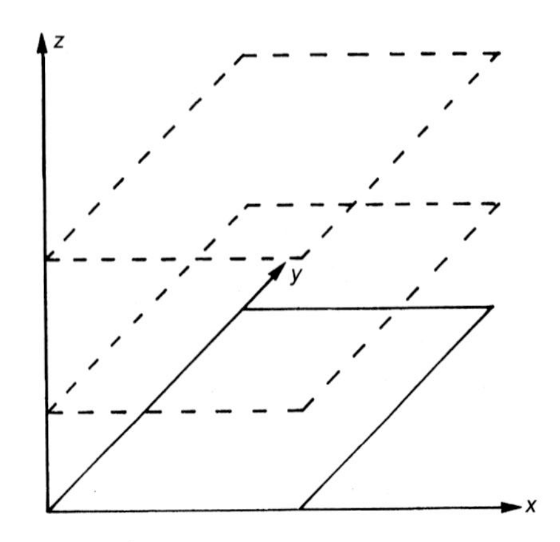

Figure 2 In an idealized electron optical analogue of a color photograph, electrons of different energies contributing to the image formed in the field of view *xy* would be displaced in the *z* direction and the intensity variation in the volume *xyz* would then be recorded.

perpendicular to the long dimension of the slit by amounts determined by the dispersion of the analyzer. A simple two dimensional recording device, such as a photographic plate, can then be used to record simultaneously the spectra from points lying along any line previously selected in the image. In the energy selecting microscope the electrons forming the image are initially dispersed into their various energy loss components by means of an analyzer, and by the use of an aperture and subsequent focusing or scanning system, one of the loss components is used to reform the image of the specimen. This image is composed of electrons with a mean energy corresponding to the selected loss component and a spectral width, usually of the order of one or two electron volts, given by the product of the dispersion of the analyzer and the width of the selecting aperture. The energy selecting microscope therefore produces the electron optical analogue of a monochrome photograph.

2. ENERGY ANALYZING AND SELECTING DEVICES

Since Wien (1897) first recognized that an energy analyzer could be designed by employing crossed electric and magnetic fields and Leithäuser (1904) constructed the first practical electron spectrometer, based on the principle of dispersion by a magnetic field, a large number and variety of electron energy analyzers have been reported in the literature (for a review see Klemperer, 1965). It is not the purpose of this article to present a comprehensive review of the instrumentation and electron optical properties of energy analyzers, but rather to consider only those analyzers currently employed in energy analyzing and energy selecting microscopes. The main restrictions on the choice of analyzer used are that the energy resolution should be of the order of 1 eV or less, and that the analyzer acceptance aperture should conform to the requirements of the imaging system of the electron microscope. The devices which meet these requirements and which have been incorporated in existing energy analyzing and energy selecting electron microscopes are:

(a) the cylindrical electrostatic analyzer;
(b) the cylindrical magnetic analyzer;
(c) the electrostatic mirror-magnetic prism analyzer.

Of these, the first has been successfully employed in both energy analyzing and energy selecting microscopes, operating at beam potentials in the range 40 to 100 kV. For high voltage electron microscopes, however, the electrostatic system presents considerable high voltage insulation problems and for this reason the cylindrical magnetic analyzer is a more suitable device to use. The third system has been used in an energy selecting electron microscope operating in the beam potential range 40 kV to 100 kV. A description of the electron optical properties of these analyzers now follows.

2.1 The Cylindrical Electrostatic Analyzer

Möllenstedt (1949, 1952) first demonstrated that an electrostatic saddle-field lens of two dimensional symmetry can be used as a high resolution electron velocity analyzer. His work has stimulated a number of theoretical papers on the electron optical properties of this system (Archard, 1954; Dietrich, 1958; Laudet, 1953; Lenz, 1953; Lippert, 1955; Metherell, 1967a; Metherell & Cook, 1972; Metherell & Whelan, 1965, 1966; Septier, 1954; Waters, 1956).

The energy resolution ΔE_r of a dispersive system, where ΔE_r is the energy difference between two just resolvable losses, is in general a function of the beam potential E of the incident electrons. For this reason it is important, when comparing the resolving powers of different devices, to compare their *specific* resolving powers, defined as $E/\Delta E_r$.

Metherell and Cook (1972) have shown that under optimum operating conditions a specific energy resolution $(\Delta E/E_r)$ of the order 10^{-7} can theoretically be achieved with the Möllenstedt system. It is doubtful that

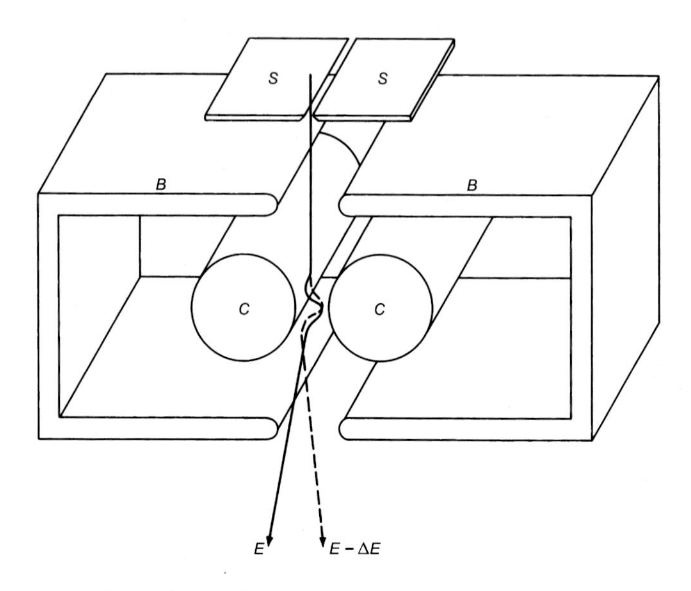

Figure 3 A view of the cylindrical electrostatic analyzer.

any other analyzer approaches this value for the specific resolution. Furthermore if a cylindrically symmetric system is employed, as opposed to the rotationally symmetric lens originally used by Möllenstedt (1949), the analyzer entrance aperture can be made in the form of a fine slit. For these reasons the cylindrical electrostatic lens is eminently suitable for performing energy analyses in the electron microscope.

A cross-sectional view of the cylindrical lens is shown in Fig. 3. A fine slit S, of width of the order of a few microns and length of the order of a centimeter, apertures the incoming electron beam. The slit S is aligned with its long dimension parallel to the axes of the cylindrical electrodes C. The apertured electrons pass into a box-shaped electrode system B at anode (usually earth) potential and then through regions of high chromatic aberration near the two cylindrical electrodes C which are biased at cathode potential. The potential distribution in the region of the electrodes C is shown schematically in Fig. 4 and the point S in this diagram is the saddle point of the system of equipotential surfaces. Provided the slit S (Fig. 3) is placed in a suitable off-axis position (see Fig. 5) electrons of different energies are dispersed, as shown schematically in Figs. 3 and 5, and the energy loss spectrum can then be recorded by placing a photographic plate, or some other suitable recording device, on the exit side of the lens.

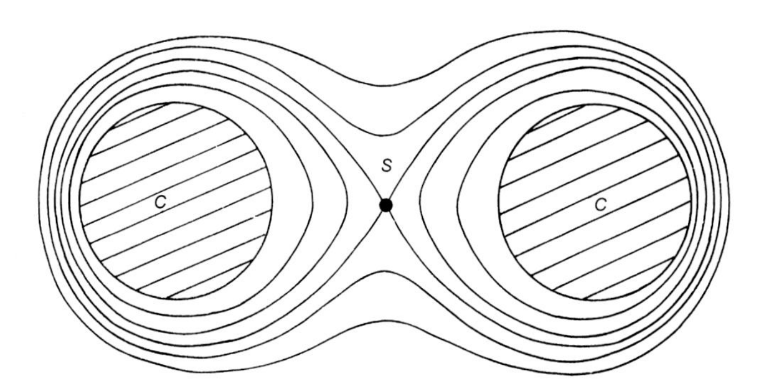

Figure 4 A cross-section of the cylindrical electrodes showing the potential surfaces in the region of the saddle point *S*.

A study of the electron optical properties of this system requires a knowledge of the potential distribution in the region between the electrodes C and B (Fig. 3). It is not possible to obtain an analytical expression for the potential distribution in a practical system with an electrode geometry such as that shown in Fig. 3. Various authors have derived expressions for the potential function in idealized electrode systems which approximate to those of practical interest. A particularly simple system, called the line-charge model of the analyzer, has been devised by Metherell and Whelan (1966). The model consists of two parallel and infinitely long line charges of charge density q per unit length (Fig. 6) placed equidistant from two earthed parallel plates AB and $A'B'$ both of infinite extent. The potential function $V(x, y)$ for this system is given by

$$V(x,y) = -q\left\{\ln\left[\frac{\cosh(\pi(x-D)/d) - \sin(\pi y/d)}{\cosh(\pi(x-D)/d) + \sin(\pi y/d)}\right] + \ln\left[\frac{\cosh(\pi(x+D)/d) - \sin(\pi y/d)}{\cosh(\pi(x+D)/d) + \sin(\pi y/d)}\right]\right\}$$

Suitable equipotential surfaces can be chosen to simulate the cylindrical electrodes of the analyzer shown in Fig. 3. Although the line-charge model does not possess equipotential surfaces which are exactly circular in cross-section and also neglects field penetration through the entrance and exit slots of the earthed electrode B of Fig. 3, Metherell and Whelan (1966) and Metherell (1967a) have shown that it represents to a good approximation the practical electrode system of Fig. 3.

The potential function of the line charge model can be used to calculate the cardinal points, aberration coefficients, etc., of the system. Our interest

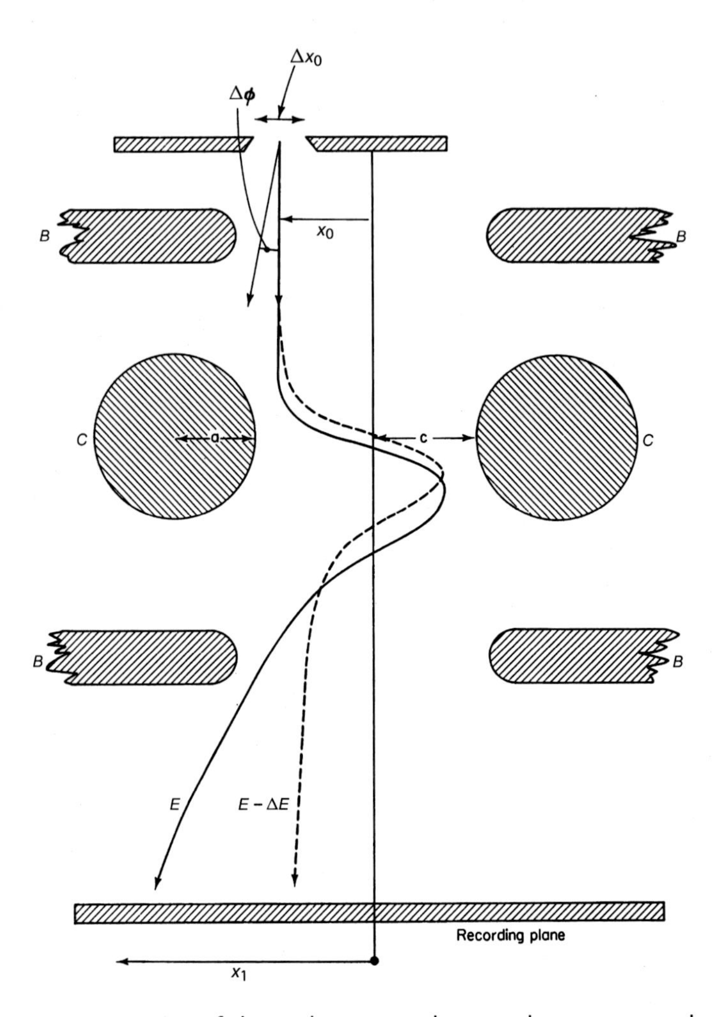

Figure 5 A cross-section of the analyzer. x_0 and x_1 are the entrance and exit co-ordinates of an electron beam passing through an entrance slit of width Δx_0.

however lies in the dispersive properties of the lens, rather than its imaging properties and the behavior of the system as an electron velocity analyzer can be best understood by examining the dispersion and resolution in terms of a plot of the entrance and exit co-ordinates x_0 and x_1 (Fig. 5) of a monoenergetic electron beam passing through the lens. An example of the x_0, x_1 curve obtained from the line-charge model is shown in Fig. 7. This curve was obtained by numerical integration of the equations of motion of an electron which was assumed to enter the system parallel to the y-axis (Fig. 6) with a beam potential of 100 kV. The units of x_0 and x_1 refer to

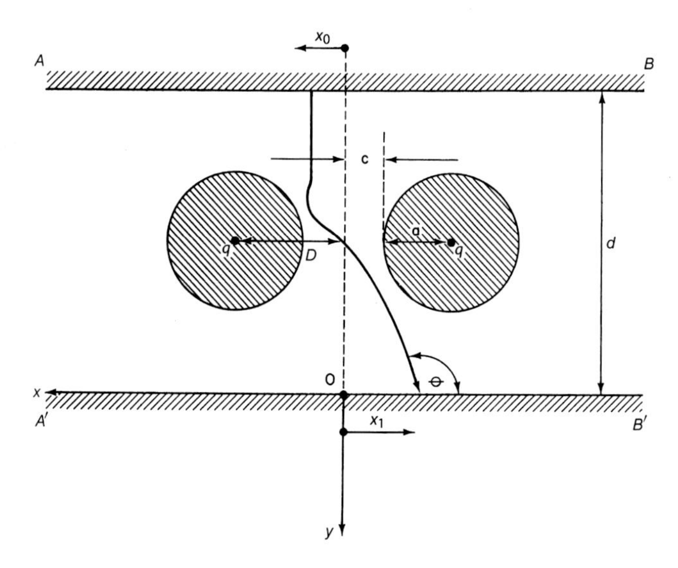

Figure 6 The line charge model of the analyzer. *AB* and *A′B′* are parallel earthed plates of infinite extent in the *x* and *y* directions. Line charges of charge density *q* per unit length intersect the plane of the diagram at the points $x = \pm D$ and $y = -0.5d$.

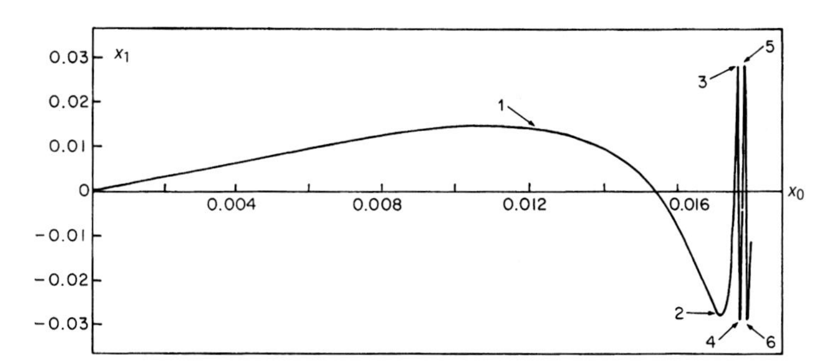

Figure 7 The x_0, x_1 curve calculated from the line charge model with c/a = 0.07 and $d =$ 1.0. The curve has been drawn schematically after the extremum 3 due to the rapidity with which x_1 varies with x_0 after this point.

$d = 1.0$, x_1 being measured in the plane $y = 0$, and the value of c/a assumed was 0.07. The curve has been drawn schematically after the second maximum (point 3 in Fig. 7) due to the rapidity with which x_1 varies after this point. The trajectories of the electrons arriving at points x_1 corresponding to the maxima and minima 1, 2, 3, and 4 of the x_0, x_1 curve are shown in Figs. 8A, B, C, and D. The extrema 1, 2, 3, etc. (Fig. 7) correspond to caustic envelopes of the system, and the trajectories of Fig. 8 are tangen-

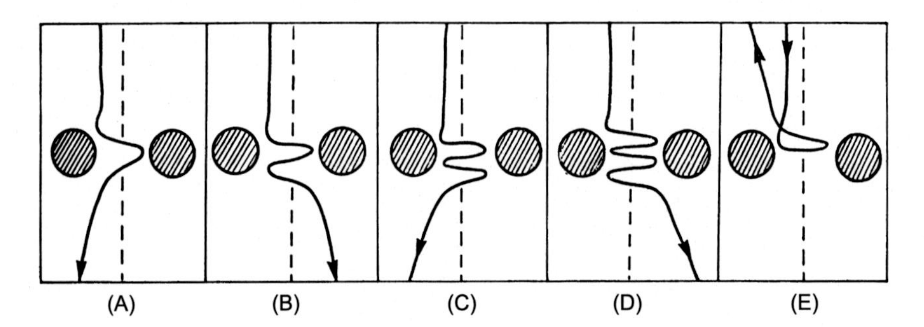

Figure 8 Trajectories of electrons arriving at points x_1 corresponding to the extrema 1, 2, 3, and 4 of Fig. 7 are shown schematically in (A), (B), (C), and (D) respectively. The trajectory of an electron reflected by the system is shown in (E).

Table 1 Definitions of the four lens classes

Class	Sign of x_1	Sign of $\cos\theta$	Approx. range of c/a
I	−	−	$\infty \to 0.32$
II	−	+	$0.32 \to 0.19$
III	+	+	$0.19 \to 0.049$
IV	+	−	$0.049 \to 0.034$

tial to these caustics at $y = 0$. In the absence of space charge effects, stray fields etc., an infinite number of extrema occur in the x_0, x_1 curve. The extremum $n = \infty$ corresponds to the case in which the electron enters the region between the inner electrodes and passes through the saddle point in a direction perpendicular to the axes of the two cylindrical electrodes. In this situation the electron becomes trapped by the lens and oscillates backwards and forwards through the saddle point. If x_0 is increased beyond this point the system behaves as a mirror and the trajectory of an electron reflected back towards the entrance side of the analyzer shown in Fig. 8E.

The most important design factor controlling the behavior of the cylindrical lens is the value of the parameter c/a (Figs. 5 and 6). The general form of the x_0, x_1 curve depends on c/a and it is found that four different types of x_0, x_1 curve are produced if c/a is varied. It is therefore convenient to divide the analyzer into four classes, depending on the sign of x_1 and $\cos\theta$ (Fig. 6). It is assumed that the electron is incident at the plane $y = -1.0$ at the positive value of x_0 corresponding to the first maximum or minimum value of x_1, measured in the plane $y = 0$. θ is the angle which the trajectory tangential to the first caustic envelope at $y = 0$ makes with the plane $y = 0$. The definitions of the four lens classes are given in Table 1, together with the approximate range of c/a values for each class. The range

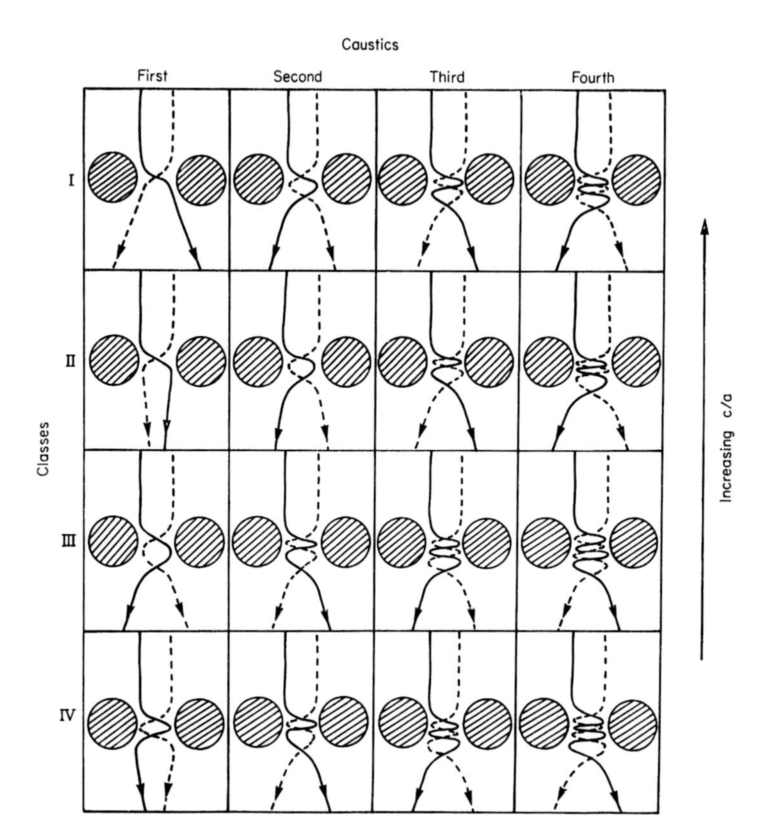

Figure 9 Electron trajectories, drawn schematically for the sake of clarity, tangential to the first four pairs of caustic envelopes at $y = 0$ (Fig. 6) in the four different lens classes.

of c/a from 0 to ~0.03 has been omitted from this table as it is convenient to postpone the discussion about this range until values of c/a > 0.03 have been considered. The electron trajectories tangential to the first four pairs of caustic envelopes at $y = 0$ in the different lens classes are shown in Fig. 9. These trajectories have been drawn schematically for the sake of clarity. Examples of the x_0, x_1 curves for the four classes of lens are shown in Fig. 10 and for each lens class three sets of curves are given corresponding to different positions of the plane in which x_1 is measured. It is worth noting that for lens classes II and IV the effect of projection below the plane $y = 0$ causes the first extremum of the x_0, x_1 curve to disappear. If the maximum and minimum values of x_1 in the x_0, x_1 curve are plotted as a function of y, the curves so obtained will give the caustic envelopes of the system, and these are shown schematically in Figs. 11A and B for the four different lens classes. It is clear from Fig. 11B that the disappearance of the first

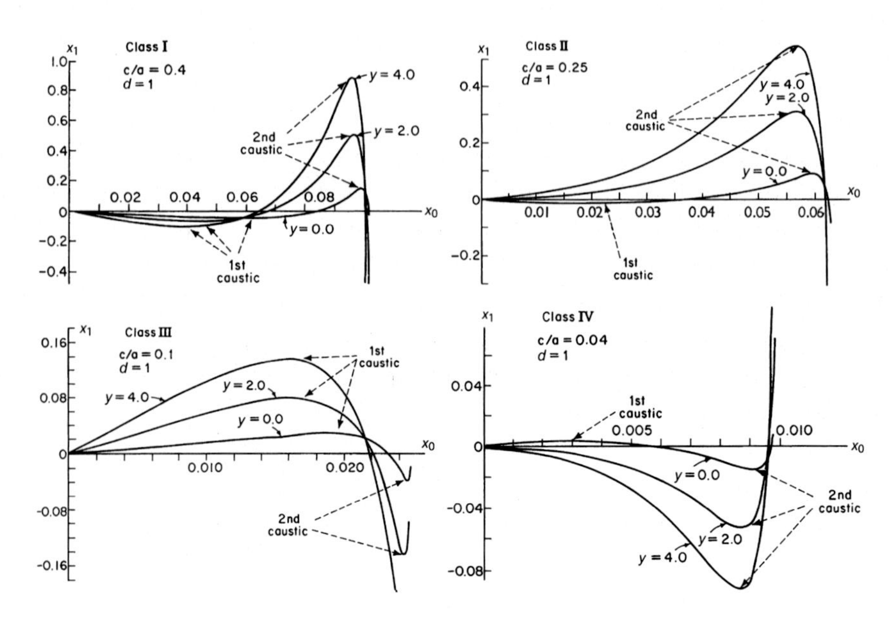

Figure 10 The x_0, x_1 curves for the different lens classes. The projection distance y, at which x_1 is measured, is indicated for each curve.

extremum in the x_0, x_1 curves for lens classes II and IV corresponds to a disappearance of the first pair of caustic envelopes with projection in these classes.

If the gap between the two cylindrical electrodes of the analyzer is gradually reduced Table 1 shows that the system passes from class I to class II to class III and then to class IV. If the value of c/a is further reduced the lens returns to class I and then passes again to classes II, III, and IV (Fig. 12). Still further reductions in c/a cause the system to cycle from one class to another in the order just given. The range of c/a values for which a particular lens class is operative decreases rapidly after the first few cycles as can be seen from Fig. 12. The reason for this cycling is indicated in Fig. 13 which shows the trajectories tangential to the first caustic envelope at $y = 0$ when the cycle of classes is repeated for the first time. It is apparent that the oscillations of the electrons in the region of the saddle point responsible for the formation of higher order caustics is also responsible for the cycling of the lens from class to class as c/a is decreased. As there is no theoretical limit to the number of oscillations an electron can make in the saddle point region, the ranges of c/a values corresponding to a particular class of lens is infinite.

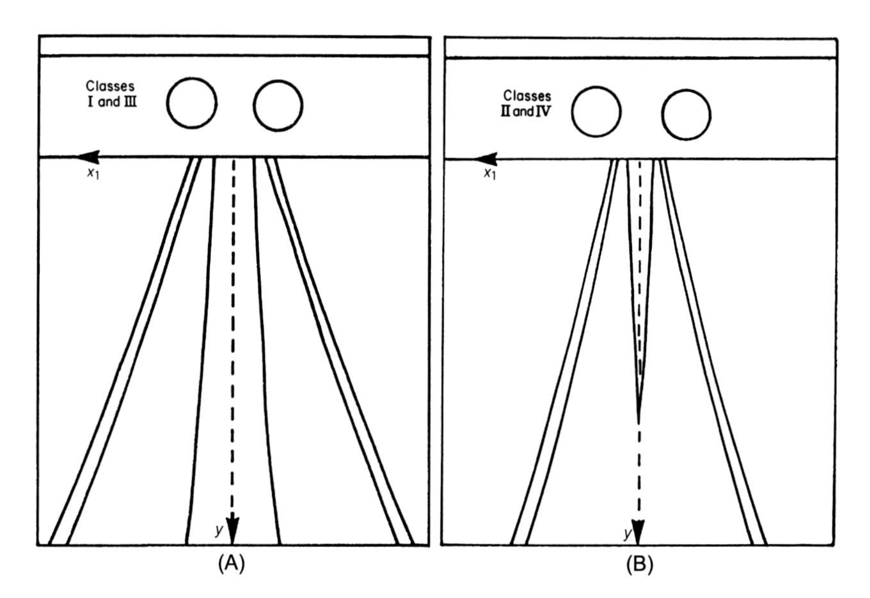

Figure 11 Behavior of the caustic envelopes produced by the cylindrical lens. In (A) the innermost caustic envelope produced by a lens of class I or III does not disappear whereas in (B) the innermost caustic envelope disappears. This latter behavior occurs with lens classes II and IV only.

The factors controlling the dispersion of the analyzer can be readily understood by reference to the x_0, x_1 curves for electrons of slightly different energies. These curves, in the region of the first and second extrema are shown schematically in Fig. 14 for electrons of energies E (full curve) and $E - \Delta E$ (broken curve). Electrons entering the analyzer at some point x_0 arrive at the recording plane separated by an amount Δx_1 (Fig. 14). The dispersion of the analyzer ($|\partial x_1/\partial E|$) is the quantity $|\Delta x_1/\Delta E|$ taken in the limit $\Delta E \rightarrow 0$, and the dependence of $\partial x_1/\partial E$ on x_0 for various positions of the recording plane is shown in Fig. 15. The vertical arrows numbered 1 and 2 indicate the values of x_0 corresponding to the first and second extrema of the x_0, x_1 curves (Fig. 10). The main point worth noting is that the dispersion has non-zero values at the first extrema of the x_0, x_1 curves for lens classes I and III, but is zero for the second, third and successive extrema. This is also true for lens classes II and IV provided that the recording plane intersects the two innermost caustic envelopes produced by these systems. If, however, the recording plane lies below the points of intersection of the two innermost caustic envelopes of lens classes II and IV (Fig. 11) the dispersion at all extrema appearing on the recording plane is zero. The reason for this behavior can best be understood by considering the x_0, x_1 curves

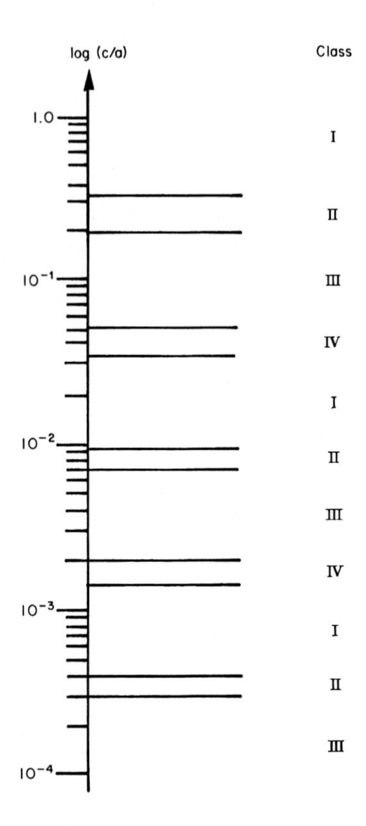

Figure 12 The range of c/a values over which each lens class is operative.

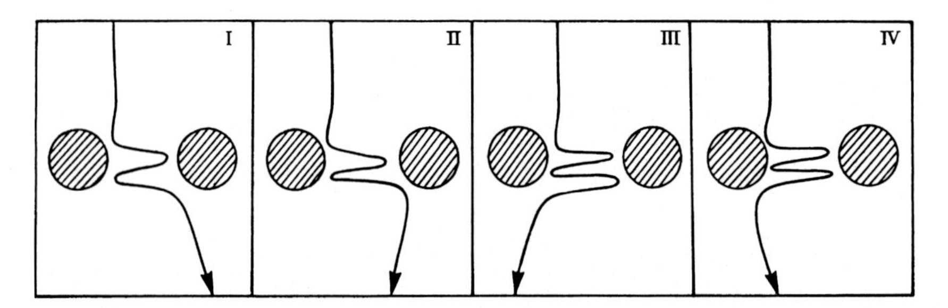

Figure 13 Caustic trajectories in the four classes of lens when the cycle of classes is repeated for the first time.

of the different lens classes for electrons of slightly different energies E and $E - \Delta E$. These curves are shown schematically in Fig. 16, where the full curves refer to an energy E and the broken curves to an energy $E - \Delta E$. It is apparent that zero dispersion occurs at the cross-over points of the curves

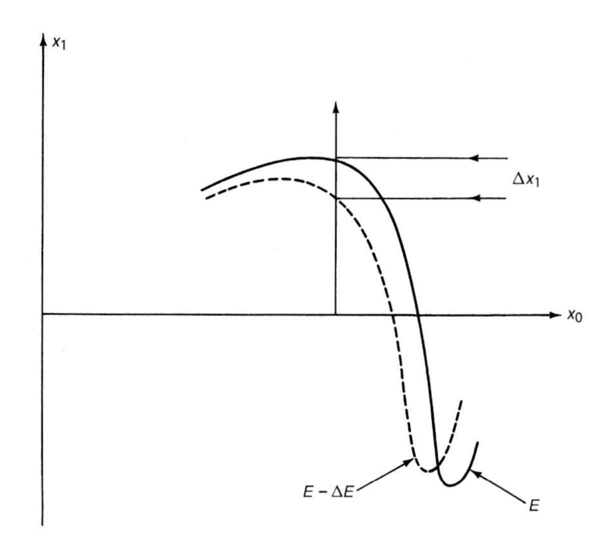

Figure 14 Schematic x_0, x_1 curves for electrons of discrete energies E and $E - \Delta E$. The dispersion is the quantity $|\Delta x_1/\Delta E|$ taken in the limit $\Delta E \to 0$.

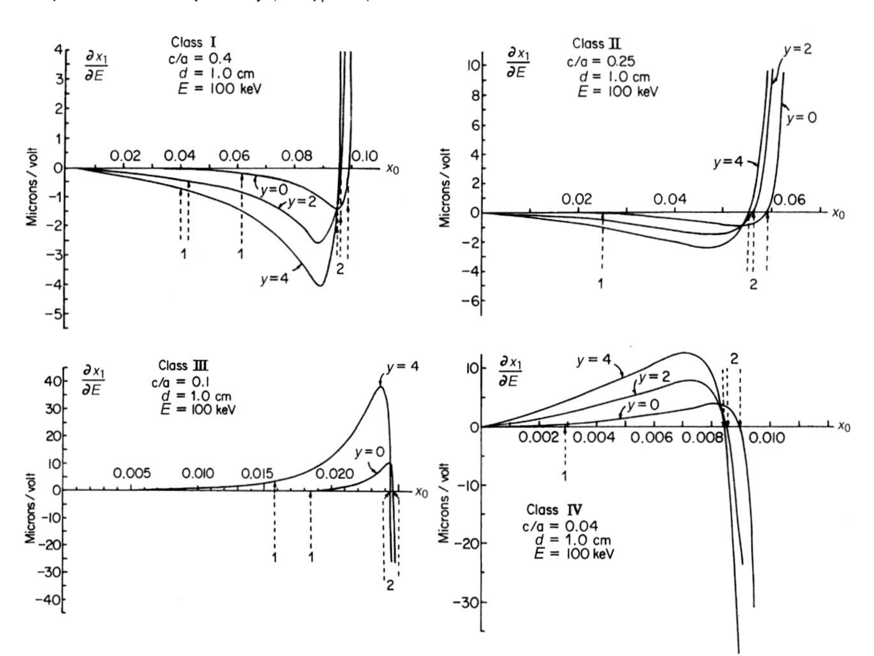

Figure 15 The behavior of the dispersion $(\partial x_1/\partial E)$ as a function of x_0 or different projection distances y. The curves in this figure were calculated from the line-charge model by assuming $d = 1$ cm (Fig. 6). The vertical arrows numbered 1 and 2 indicate the first and second caustic positions.

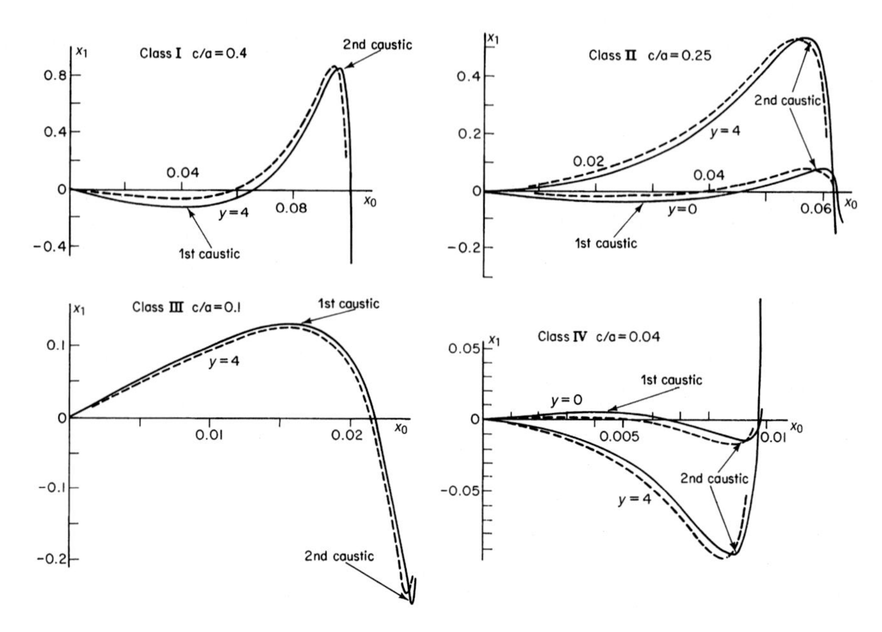

Figure 16 The x_0, x_1 curves for the four classes of lens for electrons of energy E and $E - \Delta E$. The full curves were calculated from the line-charge model with $d = 1$ (Fig. 6) and the broken curves have been drawn schematically for the sake of clarity.

for electrons of slightly different energies. In the limit $\Delta E \to 0$, the cross-over points coincide with the second and successive extrema of the x_0, x_1 curves. This behavior can be summarized by the following statement: *the dispersion of an analyzer of any class is zero at all but the first caustic edge of the system.*

The energy resolution, for a given electrode geometry, is determined mainly by

(a) the width of the entrance slit Δx_0 (Fig. 5) and

(b) the angular divergence of the beam $\Delta\phi$ (Fig. 5).

Consider first a parallel beam ($\Delta\phi = 0$) of monoenergetic electrons of energy E passing through an entrance slit of width Δx_0. It is obvious from Fig. 17, which shows the x_0, x_1 curve in the region of the first two extrema, that the electrons arrive at the recording plane in a strip of width $\delta x_1 = (\partial x_1/\partial x_0)\Delta x_0$ to the first order of small quantities. The variation of $\partial x_1/\partial x_0$ with x_0 can be obtained directly from the x_0, x_1 curves of Fig. 10, and this variation is shown in Fig. 18 for a class III lens. At the extrema of the x_0, x_1 curve the quantity $\partial x_1/\partial x_0$ is zero and δx_1 to the first order is also zero. In general a parallel monoenergetic beam passing through an

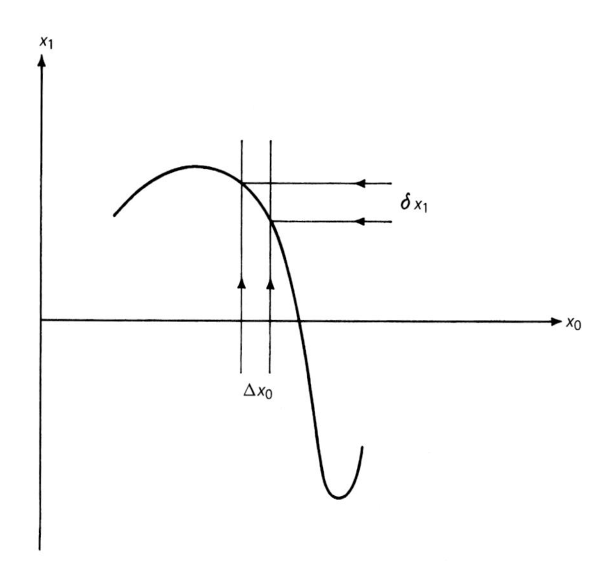

Figure 17 The x_0, x_1 curve (schematic) showing the image width δx_1 in the recording plane of an entrance slit of width Δx_0.

entrance slit of width Δx_0 will appear to have an energy spread given by

$$(\Delta E)_s = (\partial E/\partial x_1)(\partial x_1/\partial x_0)\Delta x_0 \tag{1}$$

Now consider a parallel beam containing electrons of two discrete energies E and $E - \Delta E$ incident on an entrance slit of infinitesimal width. It is obvious from Fig. 14 that the electrons arriving at the recording plane are separated by an amount Δx_1, and that electrons with an energy difference ΔE are resolved if $\delta x_1 \leqslant \Delta x_1$. Eq. (1) therefore gives the energy resolution limited by slit width.

A similar consideration shows that the resolution limited by the angular divergence $\Delta\phi$ of the beam is given by

$$(\Delta E)_\phi = (\partial E/\partial x_1)(\partial x_1/\partial\phi)\Delta\phi \tag{2}$$

Metherell and Whelan (1966) have shown that $(\Delta E)_\phi$ and $(\Delta E)_s$ are of the same order of magnitude when $\Delta\phi \sim 10^{-4}$ rad and $\Delta x_0 \sim 1$ μm. In practice slit widths ~5 μm are employed with this type of instrument, and when used as an analyzer in the energy analyzing electron microscope, the entrance aperture of the Möllenstedt system lies in the final image plane of the electron microscope. The use of an objective aperture to form the image of a specimen means that only those electrons scattered within the

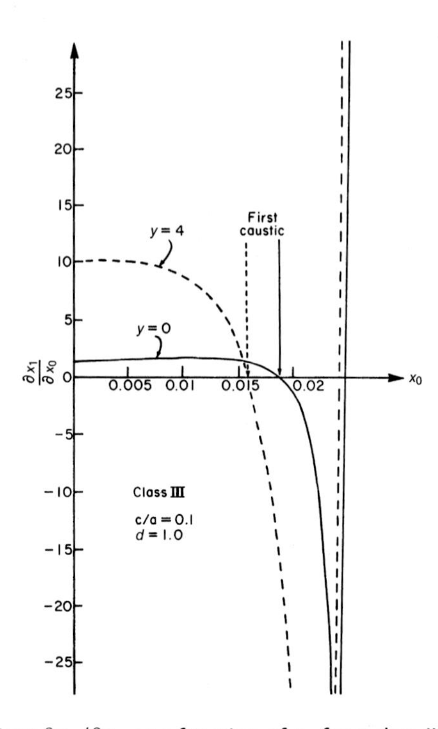

Figure 18 The gradient $\partial x_1/\partial x_0$ as a function of x_0 for a class III lens. The curve was calculated from the line-charge model with $d = 1$ (Fig. 6).

solid angle subtended by the objective aperture at a point on the bottom surface of the specimen contribute to the intensity arriving at the corresponding point in the image plane. The objective aperture semi-angle θ usually employed in 100 kV microscopes is in the range 10^{-3} to 10^{-2} rad. The electrons contributing to the intensity at a given point in the image plane therefore lie within a cone of semi-angle $\Delta\phi = \theta/M$, where M is the magnification. The value of M is typically 20,000×, which gives values of $\Delta\phi$ in the range 5×10^{-8} to 5×10^{-7} rad. The resolution limited by angular divergence is therefore negligible in comparison with the resolution limited by slit width when the analyzer is used to examine the energy spectra of electrons forming the image of a specimen in the electron microscope. In the discussion that follows the resolution $(\Delta E)_\phi$ given by Eq. (2) will be neglected and the total resolution will be assumed to be given by $(\Delta E)_s$.

Eq. (1) shows that high resolution $(E/\Delta E)_r$ is obtained when the dispersion $(\partial x_1/\partial E)$ is large and $(\partial x_1/\partial x_0)$ is small. The operating position of the analyzer entrance slit is that value of x_0 corresponding to the first extremum in the x_0, x_1 curve. With a class II or IV lens, the recording plane

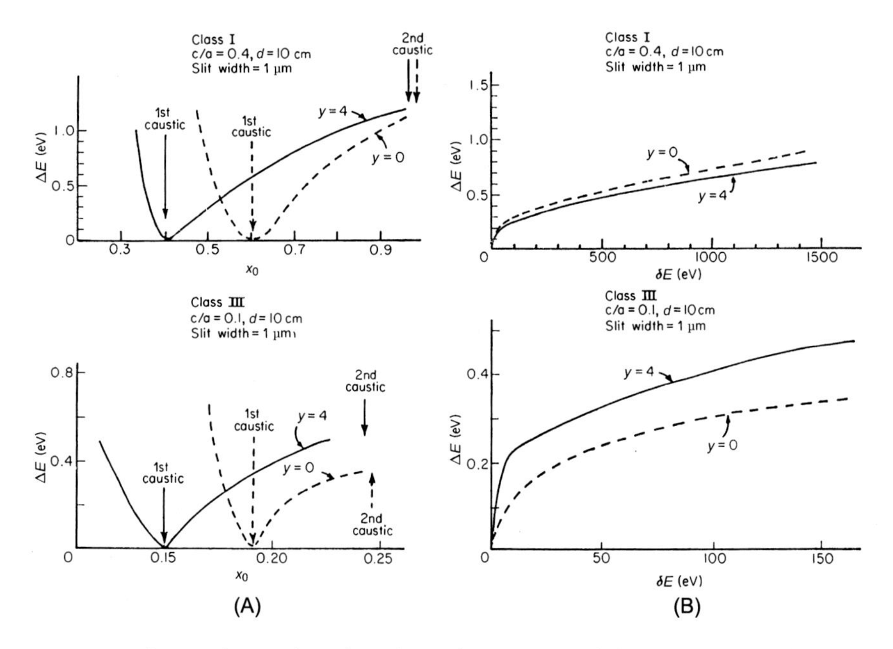

Figure 19 The resolution ΔE_r plotted as a function (A) of slit position x_0 and (B) energy loss δE for projection distances $y = 0$ and 4. These curves were calculated from the line-charge model by assuming $d = 10$ cm and an incident beam potential of 100 kV.

does not in general intersect the innermost caustic envelope and since the dispersion is zero at the second and successive caustic edges, a lens of class II or IV is unsuitable for use as a velocity analyzer. The position of the recording plane can of course be adjusted so that intersection with the first caustic edge produced by the system occurs. It is, however, desirable to have high dispersion as well as good resolution, and Fig. 15 shows that higher dispersion can be obtained with a lens of class I or III than with a lens of class II or IV if the condition that the recording plane intersects the first caustic edge of the system is satisfied.

The quantities $(\partial E/\partial x_1)$ and $(\partial x_1/\partial x_0)$ appearing in the expression for $(\Delta E)_s$ given in Eq. (1) are functions of x_0, and hence the resolution of an analyzer of given electrode geometry also depends on x_0. The energy resolution ΔE_r, assuming a beam potential of 100 kV and an entrance slit of width 1 μm, at different recording plane levels for two analyzers of different c/a values, corresponding to classes I and III, is given in Fig. 19A. This figure shows how the resolution depends on the entrance slit position x_0. In practice the slit is positioned close to the first caustic edge produced by the zero loss electrons and for this position the resolution is a minimum.

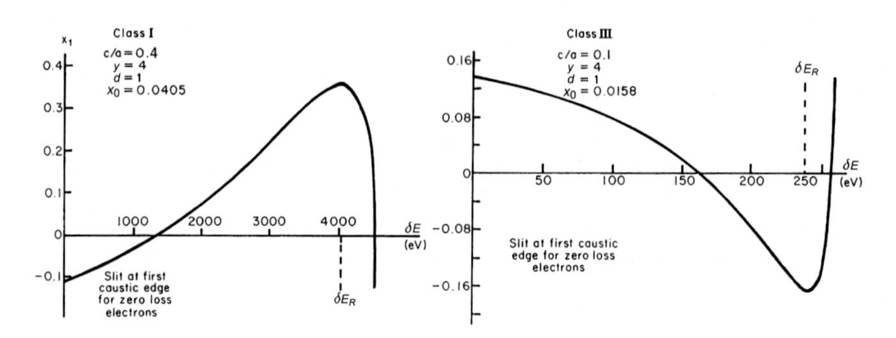

Figure 20 Calibration curves for class I and III analyzer calculated from the line-charge model. The incident beam potential is assumed to be 100 kV and the unit of x_1 refers to $d = 1.0$ (Fig. 6).

For a fixed slit position the resolution as a function of energy loss δE is not constant however, as inspection of Eq. (1) and Fig. 16 shows. The variation of ΔE_r with δE for a fixed slit position, corresponding to the first caustic edge produced by the zero loss electrons, is shown in Fig. 19B, and it is apparent that the extremely high resolution obtained at the first extremum of the x_0, x_1 curve can only be realized for energy losses lying within a few electron volts of the zero loss. It must be remembered however that at the present time the resolution for a class III analyzer is limited by the energy spread of the beam incident on the specimen, this being about 1 eV and not by the electron optical properties of the analyzer system.

In most energy loss experiments calibration of the energy loss axis of the recorded spectrum is required. The problem of calibration is eased if x_1 varies linearly with the energy loss δE for a given slit position x_0. This however is not the case for the electrostatic system, as Fig. 20 shows. The calibration curves of this figure, which refer to analyzers of classes I and III, were calculated by assuming the slit to be positioned at the first caustic edge produced by each system. In practice the calibration curve is obtained by altering the potential of the inner electrodes of the analyzer by known increments. Each voltage step simulates an energy loss and by recording these simulated losses successively on the same photographic plate the dependence of x_1 on δE can be determined experimentally.

The final factor influencing the choice of analyzer design geometry is the range of energy losses that are required to be examined. The energy loss δE is a multi-valued function of x_1 (Fig. 20) and it is therefore important that all the prominent losses in the spectrum should lie within the range $0 - \delta E_R$ where δE_R is the energy loss at which the first extremum in the δE, x_1 curve occurs. The range of energy losses of interest here (Section 1)

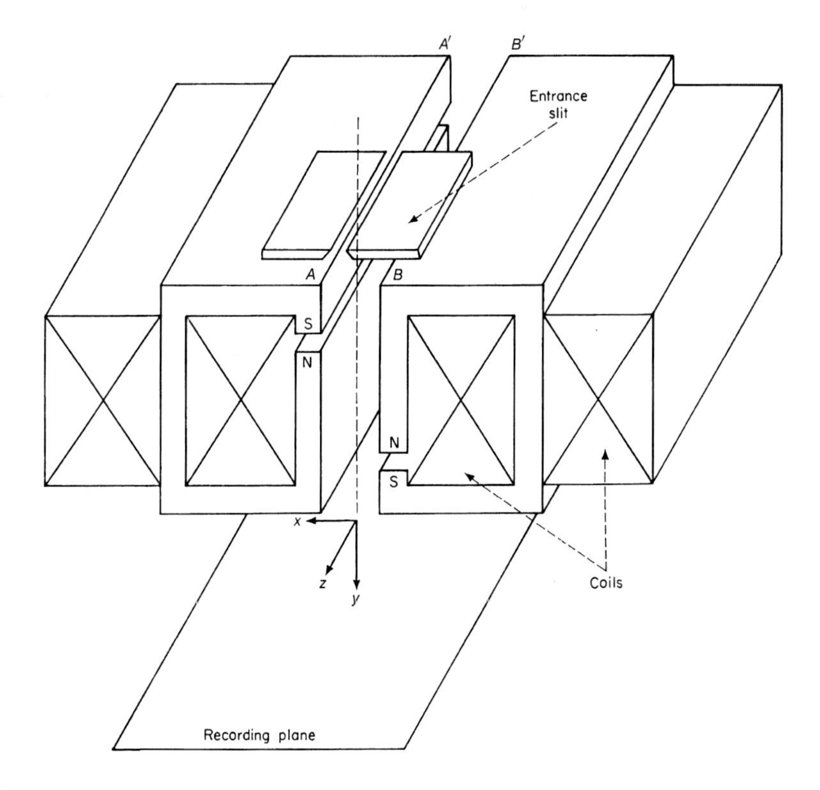

Figure 21 A view of the cylindrical magnetic analyzer.

is typically not more than about 100 eV and Fig. 20 shows that a class III analyzer operating at a beam potential of 100 kV can accommodate a range of this order of magnitude. A class III lens is therefore preferable as an analyzer since the resolution as a function of energy loss (Fig. 19) is better than that obtained with a class I lens.

2.2 The Cylindrical Magnetic Analyzer

At high accelerating voltages the problems of electrical insulation make the use of an electrostatic analyzer undesirable. It is obviously preferable to seek a purely magnetic system and so avoid these problems altogether. A suitable magnetic device, of compact form and high energy resolution, is a cylindrical system first described by Ichinokawa (1965) (see also Considine, 1970; Considine & Smith, 1968; Ichinokawa, 1968; Ichinokawa & Kamiya, 1966). This analyzer is the magnetic analogue of the electrostatic system discussed in Section 2.1 and consists of two pairs of pole pieces displaced relative to each other in the manner indicated in Fig. 21. The coils are

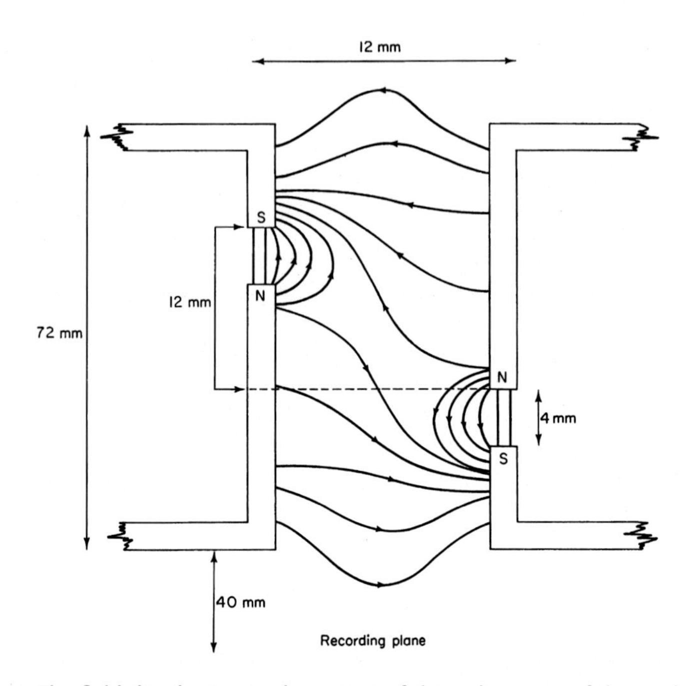

Figure 22 The field distribution in the region of the pole pieces of the analyzer. The dimensions given in this figure are those assumed by Considine (1970) in calculating the curves of Figs. 35, 36, and 38 of this article.

energized so that the magnetic polarities across the pole-piece gaps are in opposite senses. The reason for this reverse polarity and also the asymmetric arrangement of the gaps between the pole pieces is considered below. An entrance slit placed above the pole pieces apertures the incoming electron beam. The field distribution in the space between the pole pieces is shown schematically in Fig. 22 together with the dimensions of the system incorporated with the Cavendish 750 kV electron microscope (Considine, 1970; Considine & Smith, 1968). The field distribution deflects the incoming electron beam in both the xy and yz planes (Fig. 21) and if the entrance slit is placed in a suitable off-axis position, electrons of different energies are dispersed in the highly inhomogeneous fields near the pole pieces of the system. The trajectories of electrons of slightly different energies projected on the xy and yz planes (Fig. 21) are given in Figs. 23A and B. The xz projection of the system is given in Fig. 24. In this diagram the broken lines AA', BB' correspond to the pole-piece faces AA' and BB' shown in Fig. 21 and the entrance slit is represented by S_0S_0'. A parallel beam of electrons of two discrete energies E and $E - \Delta E$ incident on S_0S_0' arrive

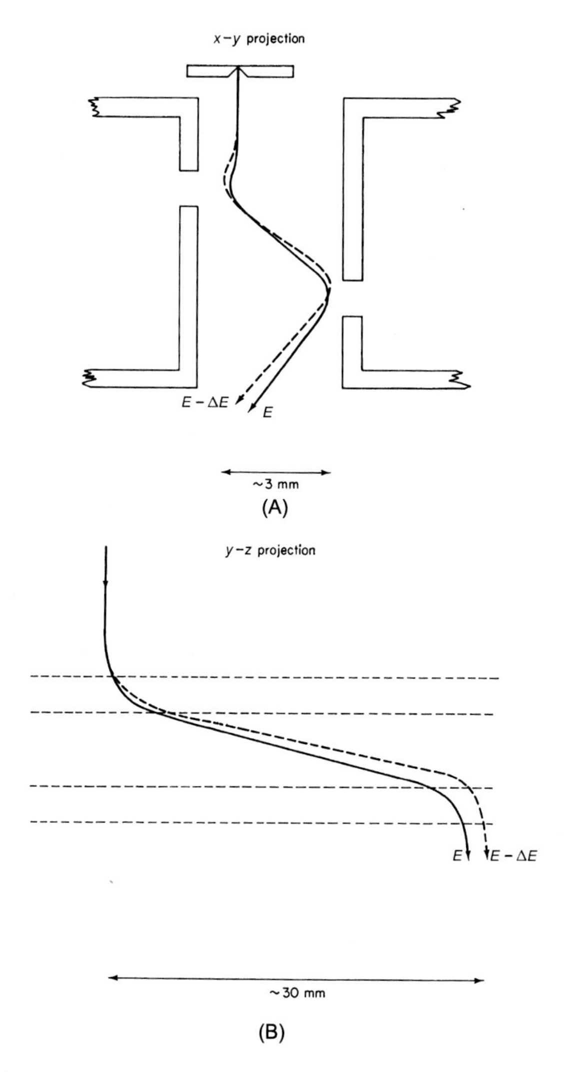

Figure 23 (A) The *xy* projections and (B) the *yz* projections of electron trajectories in the system.

at the recording plane xz along the lines S_1S_1' and S_2S_2' respectively. One of the major differences therefore between the magnetic analyzer and its electrostatic analogue is that in the latter, electrons initially parallel to the y direction suffer no deflection in a direction parallel to the axes of the cylindrical electrodes (the z direction).

The reason for the reverse polarity across the pole-piece gaps is that this mode of excitation reduces the total deflection of the electron beam in the

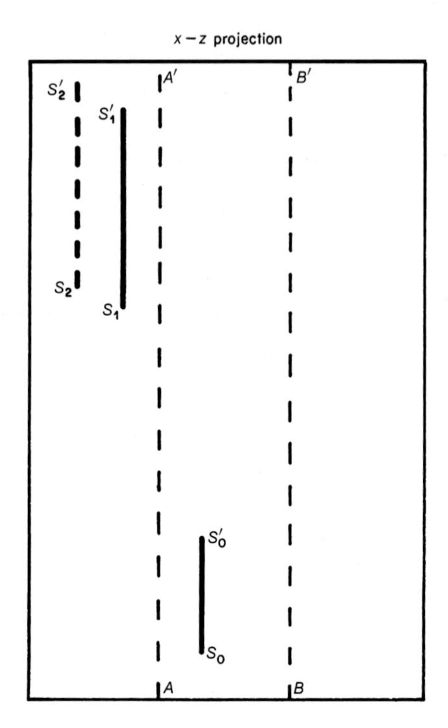

Figure 24 The *xz* projection.

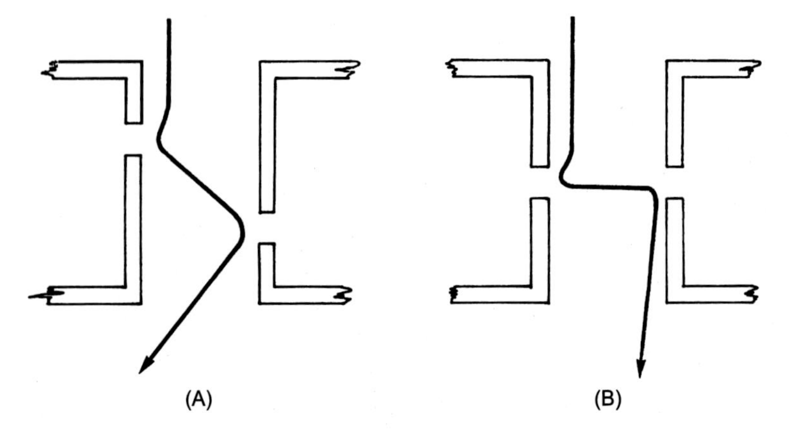

Figure 25 Electron trajectories in systems where the pole pieces are placed (A) asymmetrically and (B) symmetrically with respect to each other.

yz plane (Fig. 23B). The gaps are placed at different levels since this reduces the field strength required to deflect the beam into a trajectory such as that shown in Fig. 25A. The field strength required for the trajectory

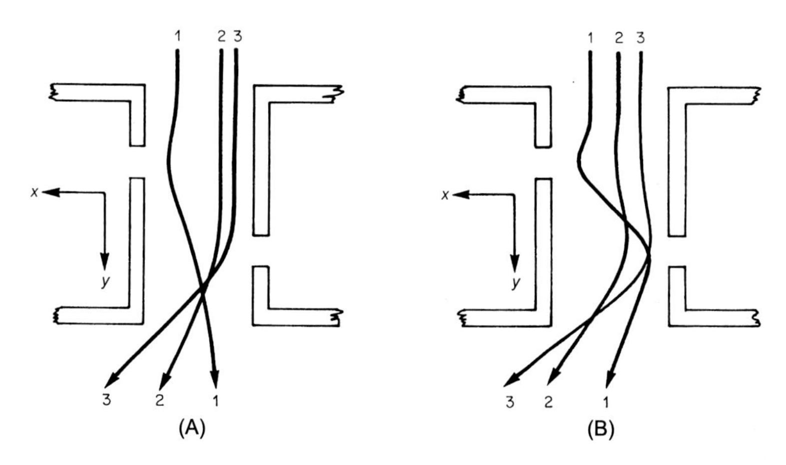

Figure 26 Electron trajectories in the system when (A) low energizing currents are used and (B) when the energizing current is high.

in the system illustrated in Fig. 25B is much higher than that required for the asymmetric system of Fig. 25A. Schematic trajectories of electrons of the same energy entering the system at different points along the x axis are given in Figs. 26A and B. The excitation of the lens is assumed low in Fig. 26A and the effect of increased excitation is shown in Fig. 26B. If the excitation is further increased electrons incident at 3 will be deflected into the pole-piece face and will be lost from the system. Still further excitation causes electrons incident at 2 also to be lost from the system. In practice the lens is operated at a sufficiently high excitation so that only those electrons entering the system in the neighborhood of trajectory I pass through the analyzer. The effect of increasing excitation on the trajectories of electrons incident at 1 (Fig. 26) is shown schematically in Fig. 27 and the dependence of the exit coordinate x_1 on the energizing current I is shown in Fig. 28. The x_1, I curve intercepts the x_1 axis at a value $x_1 = x_0$ (Fig. 27) and at a value I_c of the excitation current the beam strikes the edge of the lower pole-piece. It must be remembered that the trajectories of Fig. 27 represent the xy projection of the true paths followed by the electrons and as I is increased so the deflection suffered by the electrons in the yz plane is increased. It is worth noting that the displacement in the z direction can be reduced, or entirely eliminated by rotating the analyzer about an axis perpendicular to the yz plane. This is illustrated in Fig. 29 where the broken line indicates the trajectory projected on the yz plane for an electron incident parallel to the y direction and the full line the corresponding projection for an electron incident at an appreciable angle to the y direction.

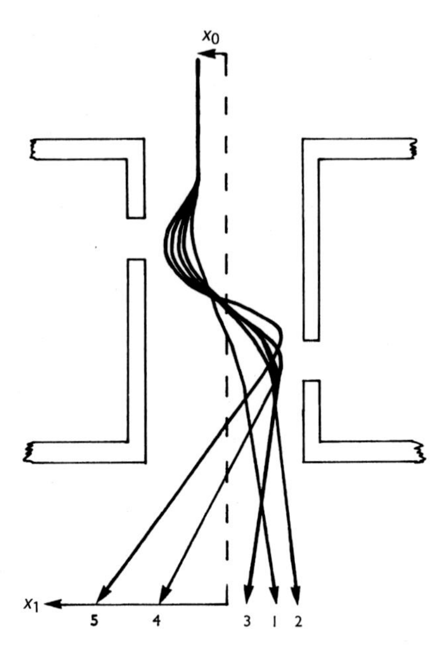

Figure 27 The effect of increasing the energizing current on the trajectory of an electron incident at x_0.

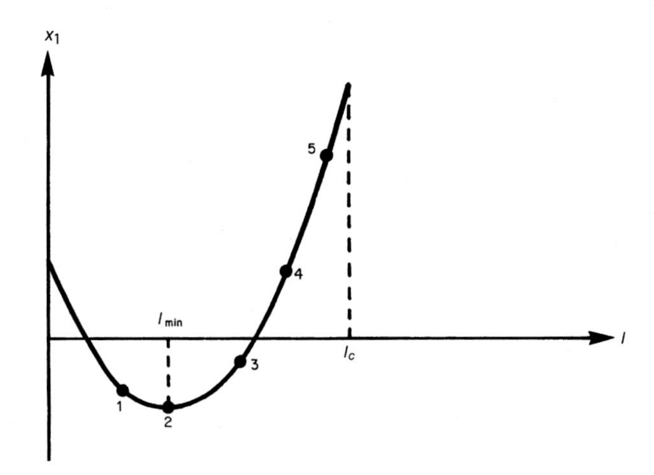

Figure 28 The variation of x_1 with energizing current I for an electron incident at x_0 (Fig. 27).

Consider now a monoenergetic beam entering the analyzer at different points x_0 (Fig. 30A) when the excitation I (Fig. 28) is less than I_{min}. The x_0, x_1 curve for this case is shown in Fig. 30B. Electrons incident at values

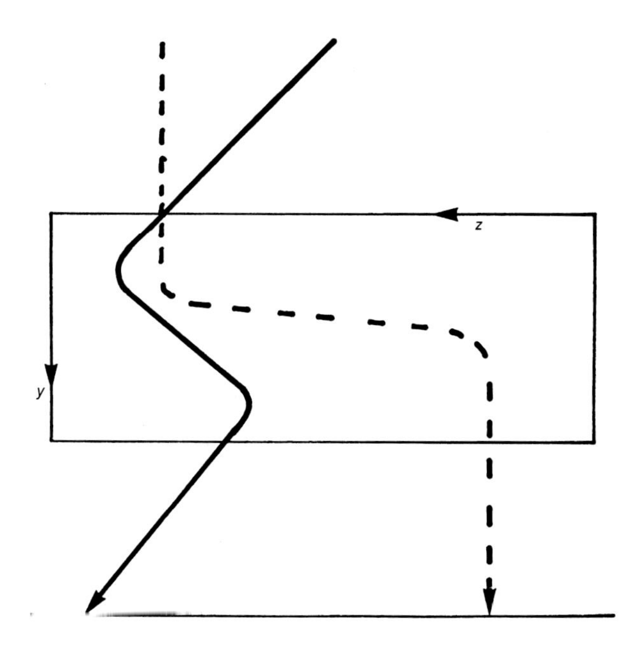

Figure 29 The *yz* projections of electron trajectories. The comparatively large deflection of the electron in the *z* direction can be reduced by allowing the electron beam to be incident at an angle to the *y* axis.

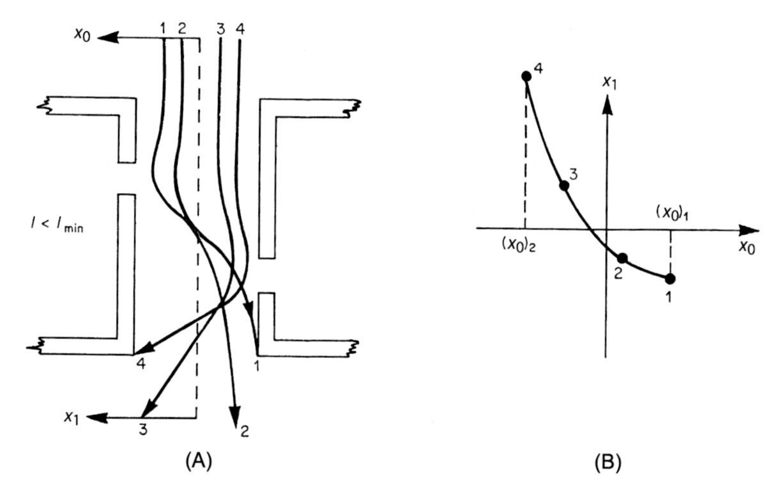

Figure 30 (A) Trajectories of electrons incident at different points x_0 when the energizing current $I < I_{min}$ (Fig. 28). (B) The x_0, x_1 curve obtained when $I < I_{min}$.

of $x_0 < (x_0)_2$ and $x_0 > (x_0)_1$ are lost from the system and the x_0, x_1 curve possesses no maxima or minima. If however I is greater than I_{min} the trajectories behave in the manner indicated in Fig. 31A and a minimum appears

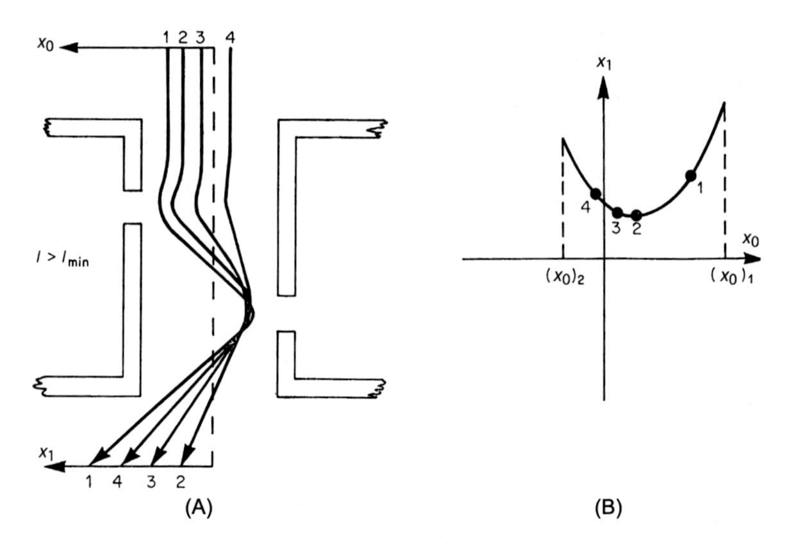

Figure 31 (A) Trajectories of electrons incident at different points x_0 when $I > I_{min}$ (Fig. 28). (B) The x_0, x_1 curve when $I > I_{min}$.

in the x_0, x_1 curve (Fig. 31B). The system therefore possesses a caustic edge and the behavior of the x_0, x_1 curve is similar to that of the electrostatic analyzer (Section 2.1). The existence of this caustic edge is the reason for the high resolving power ($E/\Delta E_r \sim 10^5$) reported in the literature for this system. The behavior of the x_0, x_1 curves for different excitation currents is illustrated in Fig. 32 where $I_4 > I_3 > I_2 > I_1 > I_{min}$. The minima of these curves become increasingly sharp as the excitation current is increased and it is apparent that the image width δx_1 (defined in Section 2.1, Fig. 17) of a parallel monoenergetic beam entering the system through an entrance slit of width Δx_0, positioned at the appropriate caustic edge, increases as the excitation increases. The resolution is given by $\Delta E_r = (|\partial E/\partial x_1|)\delta x_1$ and the behavior of δx_1 with excitation suggests therefore that the optimum excitation for energy analysis should be close to I_{min}. The dispersion $(|\partial x_1/\partial E|)$, however, is found to increase more rapidly than δx_1 with increasing excitation with the result that the smallest value for the resolution ΔE_r occurs at I_c. The effect of excitation on the dispersion is considered in more detail below.

The x_0, x_1 curves of Fig. 32 do not indicate how the displacement of the beam in the z direction is influenced by the pole-piece excitation. The effect of lens excitation on this displacement can best be illustrated by considering the trace that the point of intersection of an electron beam

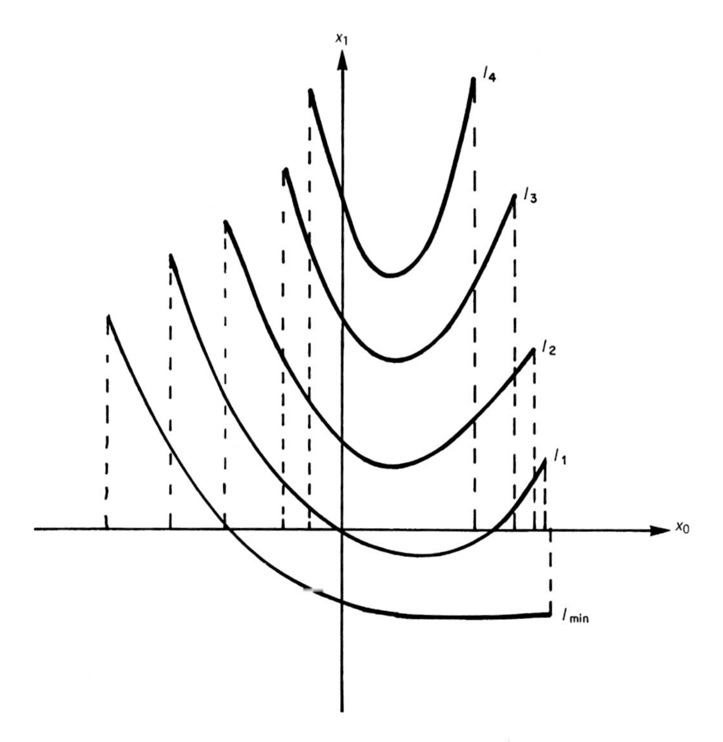

Figure 32 The x_0, x_1 curves obtained for different energizing currents ($I_4 > I_3 > I_2 > I_1 > I_{min}$).

of infinitesimal cross-section makes with the recording plane xz as the entrance position x_0 is varied. The traces obtained for excitations greater than I_{min} are given schematically in Fig. 33. The arrow on any one of the traces indicates the direction of movement of the point of intersection as x_0 is increased.

The effect of excitation on the dispersion of the analyzer can also be illustrated by considering the x_0, x_1 curve for electrons of two discrete energies E and $E - \Delta E$. The full lines of Fig. 34 refer to electrons of energy E and the broken curves to energy $E - \Delta E$. The main points of interest are that the dispersion $(\Delta x_1/\Delta E)$ increases with increasing excitation and at the caustic edge $(\Delta x_1/\Delta E)$ is non-zero. The dispersion obtained at the caustic edge of an analyzer with pole-piece dimensions given in Fig. 22, and with the recording plane at a distance 40 mm below the base of the system, is shown in Fig. 35. The dispersion was calculated (Considine, 1970) by assuming a fractional energy change $\Delta E/E = 10^{-3}$. The separation Δx_1 (Fig. 34) of electrons of energies E and $E - \Delta E$ was then found by us-

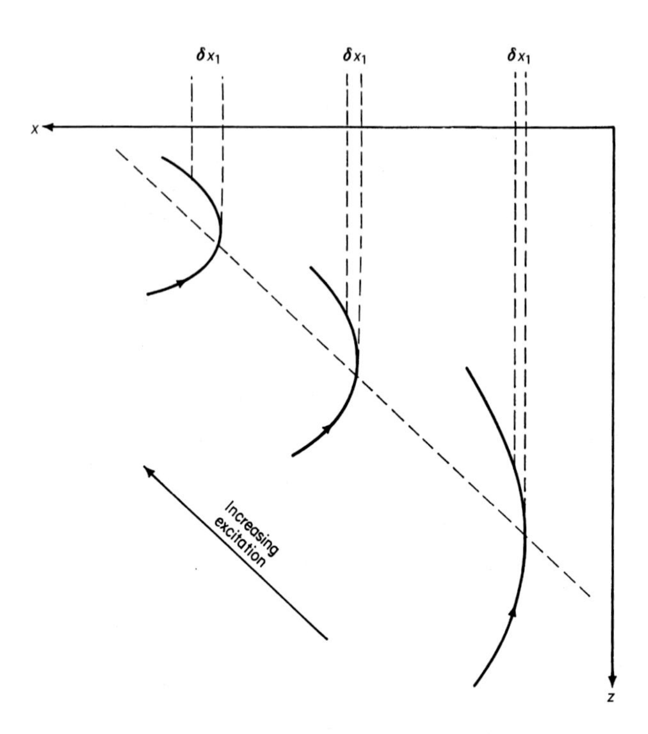

Figure 33 A schematic diagram showing the traces that would be obtained on the recording plane *xz* by illuminating an entrance slit of infinitesimal width aligned parallel to the *x* axis (Fig. 21). The image width δx_1 of an entrance slit of width Δx_0 increases as the excitation is increased.

ing numerical trajectory tracing techniques. The horizontal axis of Fig. 35 refers to an excitation parameter k defined by the relation

$$NI = kV_r^{\frac{1}{2}} \tag{3}$$

where NI is the number of ampere turns used to excite the lens and

$$V_r = V\left(1 + \frac{eV}{2m_0c^2}\right)$$

where V is the beam potential, m_0 the rest mass of the electron, e its charge and c is the velocity of light. Eq. (3) gives the excitation required to produce a trajectory of a given shape as a function of the beam potential. Also plotted as a function of k in Fig. 35 is the width δx_1 (Fig. 17) measured at the caustic edge for an entrance slit of width 15 μm. Using the expression $\Delta E_r = (|\Delta E/\Delta x_1|)\delta x_1$, the resolution as a function of k can be calculated

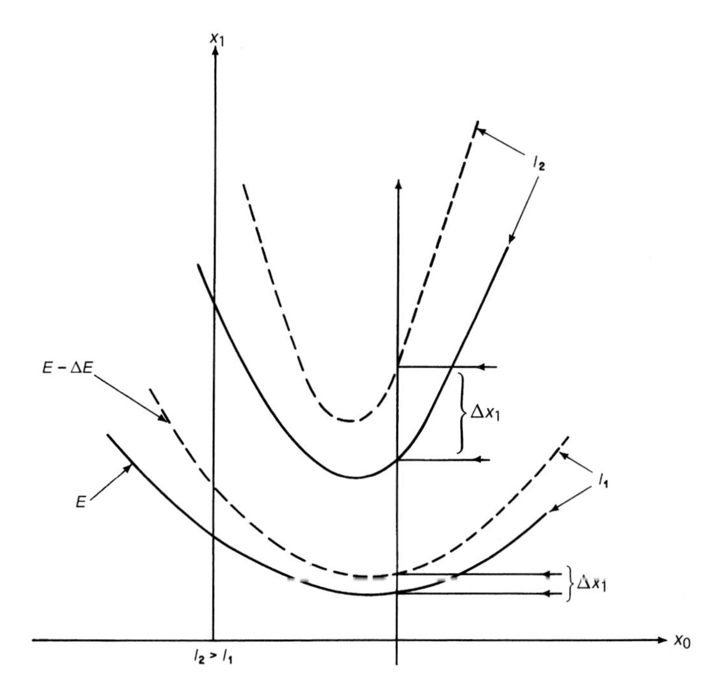

Figure 34 The x_0, x_1 curves for electrons of different energies E and $E - \Delta E$. The dispersion is the quantity $|\Delta x_1 / \Delta E|$ taken in the limit $\Delta E \to 0$.

from the information given in Fig. 35 and it is found that ΔE is very nearly independent of k. The specific resolution $(\Delta E/E)_r$ falls from 1.3×10^{-5} at $k_{\min}$ to 0.9×10^{-5} at k_c where $k_{\min}$ and k_c are the excitation parameters for currents $I_{\min}$ and I_c of Fig. 28. The dispersion for the system of Fig. 22 at excitation $k = 0.52$ amp. turns $(\text{volts})^{-\frac{1}{2}}$ as a function of the fractional energy loss $\Delta E/E$ is given in Fig. 36, which shows that the dispersion is linear for losses up to $\Delta E = 3 \times 10^{-3} E$.

It is to be noticed that the dispersion of this system is extremely small; for example at k_c, the value for a beam potential of 100 kV is only ~5 μm/V compared with ~50 μm/V for an electrostatic analyzer of similar geometry. Obviously the dispersion can be increased by increasing the dimensions of the analyzer by some constant factor and also by increasing the projection distance. If photographic techniques are to be employed for recording the energy loss spectra a dispersion of about 100 μm/V is necessary if problems associated with the limited spatial resolution of photographic emulsions are to be avoided. To achieve a dispersion of 100 μm/V with the design dimensions of Fig. 22, a projection distance ~100 cm is required which

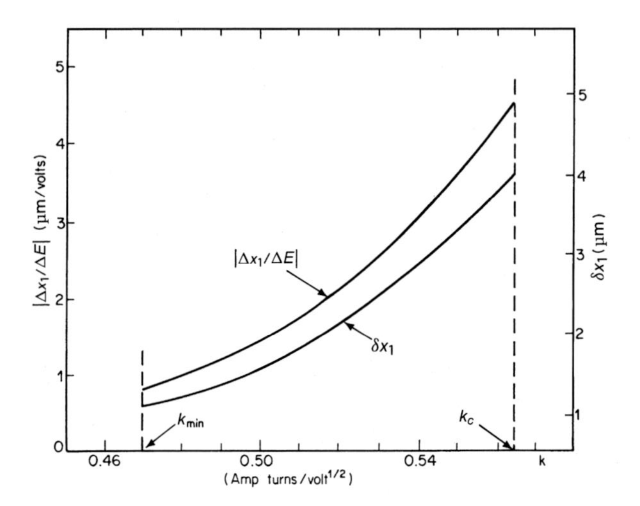

Figure 35 Variation of the dispersion $|\Delta x_1/\Delta E|$ and image width δr_1 (defined in Fig. 17) with the excitation parameter k. The quantities plotted in this figure refer to an analyzer with design dimensions given in Fig. 22. The width of the entrance aperture assumed in calculating δx_1 is 15 μm and the beam potential assumed in calculating $|\Delta x_1/\Delta E|$ is 100 kV.

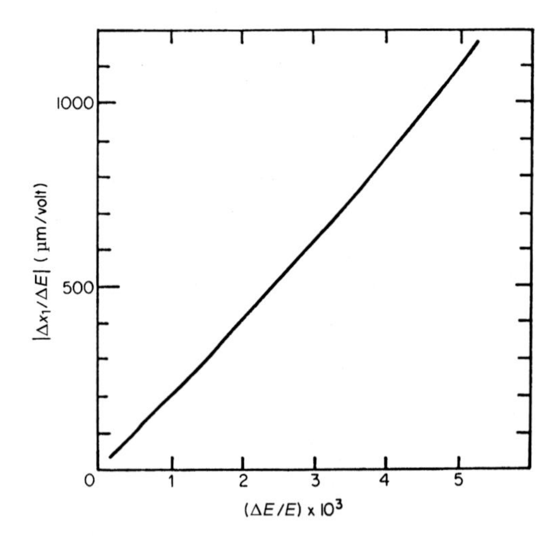

Figure 36 The dispersion $|\Delta x_1/\Delta E|$ plotted as a function of the fractional energy loss $\Delta E/E$ (after Considine, 1970).

is obviously impractical. To overcome this problem Ichinokawa (1965) mounted the analyzer between the intermediate and projector lenses of an electron microscope and used the magnification produced by the projector

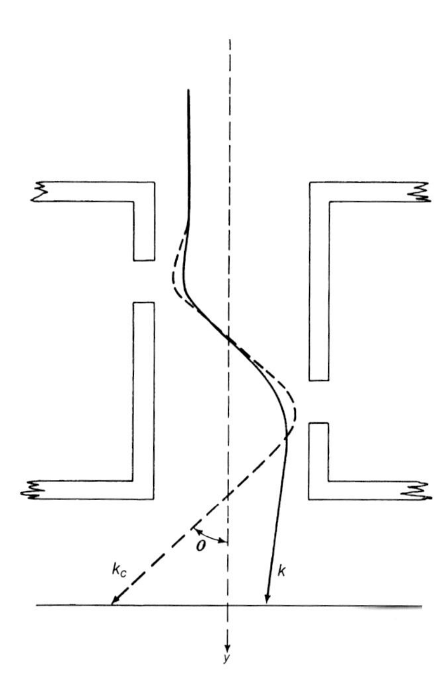

Figure 37 Electron trajectories for two excitation parameter $k = k_c$ and $k \geqslant k_{min}$.

lens to achieve a reasonable dispersion. It is however highly undesirable to position an analyzer between these two lenses, since it can interfere with the normal operation of the microscope and more importantly reduces the *spatial* resolution available for microanalysis (Section 3.1). If photographic techniques are to be used to record the spectra it is preferable to mount the analyzer in the final image plane of the microscope and employ an auxiliary lens placed below the analyzer to achieve the required magnification. Alternatively the spectra can be recorded electronically (Section 3.1) in the manner employed by Considine and Smith (1968) and Considine (1970).

In view of the small dispersion associated with the magnetic system it would seem desirable to excite the pole pieces to a value of k as close to k_c (Fig. 35) as possible. There are, however, several practical considerations to be made before the optimum excitation can be decided on. Consideration must first be given to the range of energy losses that is to be examined. Caustic trajectories for electrons of the same energy E are given in Fig. 37 for excitation parameters k, where $k \geqslant k_{min}$, and k_c. These trajectories, however, are also those for electrons of energy E (full curve) and $E - (\Delta E)_R$ (broken curve) where $(\Delta E)_R$ is the range of losses accepted by the ana-

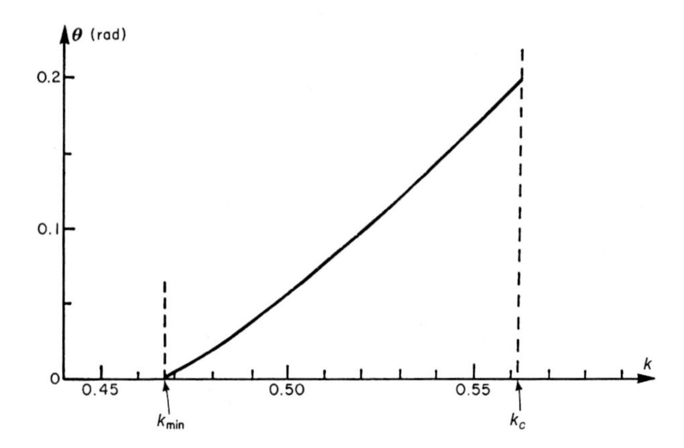

Figure 38 The angle θ (Fig. 37) plotted as a function of the excitation parameter k (after Considine, 1970).

lyzer when the excitation is k. Provided $(\Delta E)_R \ll E$, Eq. (3) shows that the largest excitation parameter (k_{max}) that can be used if the range of losses is $(\Delta E)_R$ is given by

$$k_{max} = k_c(1 - (\Delta E)_R/E)$$

The fractional range $(\Delta E)_R/E$ of losses contributing to the microscope image is usually in the region of 10^{-3}. The energy loss range is therefore an unimportant factor in deciding the optimum value of k to be used, since the equation for k_{max} given above shows that values of k close to k_c can be used if so desired. Consideration must also be given to the angle θ (Fig. 37) that the emerging beam makes with the y axis. If spectra are to be recorded photographically, it is desirable that the beam entering the auxiliary lens should make as small an angle as possible with the y axis. Similarly, if the spectra are to be recorded electronically, alignment coils placed below the analyzer (Section 3.1) are required to deflect the beam onto a detector, and the design of the deflection coils is simplified if θ is small. The angle made by the trajectory projected on the yz plane and the y axis is very small (less than $\sim 10^{-5}$ rad for $(\Delta E)_R/E = 10^{-3}$) whereas the angle made by the trajectory in the xy projection at $k = k_c$ is $\sim$0.2 rad. The variation of θ with k is given in Fig. 38 and it is seen that θ is very small when $k = k_{min}$. High dispersion and small θ are therefore conflicting requirements. If the spectra are recorded photographically, then values of k close to k_{min} can be used, provided that an auxiliary lens of sufficient magnification is employed. If, however, electronic detection is used, then it is found that yet another factor, namely the width of the energy selecting aperture placed over the detecting system, influences the choice of excitation parameter used. A full discussion of this point is given in Section 3.1.

2.3 Mirror Prism Device

The possibility of devising a filter lens which could be placed immediately after the objective lens of an electron microscope and allow normal operation of the intermediate and projector lenses to produce highly magnified energy selected images was the subject of theoretical studies by Hennequin (1960) and Paras (1961). Hennequin studied the dispersive properties of magnetic prisms whereas Paras studied a number of devices among which was a system consisting of a combined electrostatic mirror and magnetic prism. This analyzer was subsequently developed by Castaing and Henry (1962, 1963, 1964) as an integral part of an energy selecting electron microscope which is described in Section 4.1 of this article.

The filter lens consists of a double magnetic prism, employing a uniform magnetic field, and a concave electrostatic mirror biased at cathode potential (Fig. 39). Under certain conditions, which are considered below, the system possesses two sets of stigmatic points; one set $R_1R_2R_3$ (Fig. 39) is real, the other set $V_1V_2V_3$ is virtual. The reason for the existence of these points is illustrated in Figs. 40A and B. A convergent beam of electrons (Fig. 40A) initially focused at V_1, so that V_1 is a virtual stigmatic object point, is deviated by the prism so that it appears to diverge from the point V_2, which is a virtual stigmatic image point; similarly a convergent beam focused initially at V_2 appears to diverge from the point V_3. A beam of electrons diverging from R_1 (Fig. 40B) which is a real stigmatic object point is however brought to a real focus at R_2, and similarly a beam diverging from R_2 is brought to a real focus at R_3. The positions of these stigmatic points, and indeed their very existence and character, are determined by the angle of incidence which the incoming electrons make with the entrance face of the prism.

Before considering the formation of stigmatic object and image points in this system, a general analysis of the electron optical properties of the magnetic prism will be given. Consider a point object A (Figs. 41A and 41B) emitting a fine cone of electrons in the direction AP, and suppose that the mean trajectory $APP'\,A'_1\,A'_2$ makes an angle ϵ_1 (Fig. 41B) with the normal to the face DE of the prism. In the region of the uniform magnetic field enclosed by the faces CD and DE of the prism, the mean trajectory has a radius of curvature a and the electron emerges from the second face

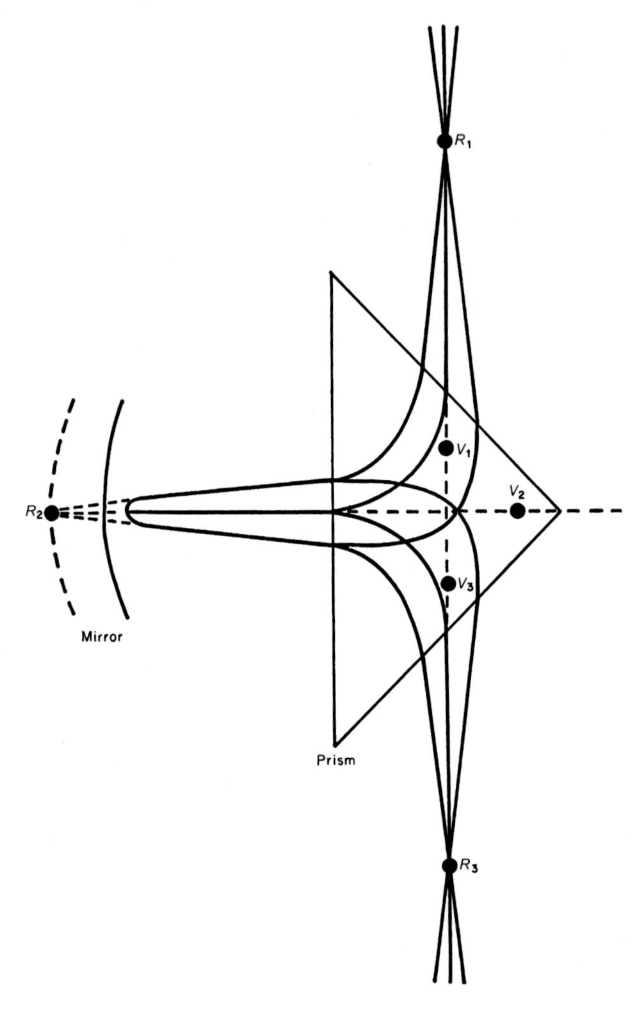

Figure 39 The double magnetic prism and electrostatic mirror. R_1, R_2, and R_3 are real stigmatic points and V_1, V_2, and V_3 are virtual stigmatic points.

CD of the prism at an angle ϵ_2, suffering a total angular deviation θ. In the general case two focal lines, A_1' and A_2', are formed in image space corresponding to the object point A. The radial focal line A_1' is parallel to the magnetic induction B and the axial focal line A_2' is perpendicular to B. The magnetic prism is therefore astigmatic and possesses two sets of cardinal points. The first set corresponds to electrons whose trajectories lie in a plane perpendicular to the magnetic induction B (the plane *CDE* of Fig. 41A). This plane is called *the first principal section* of the system. The

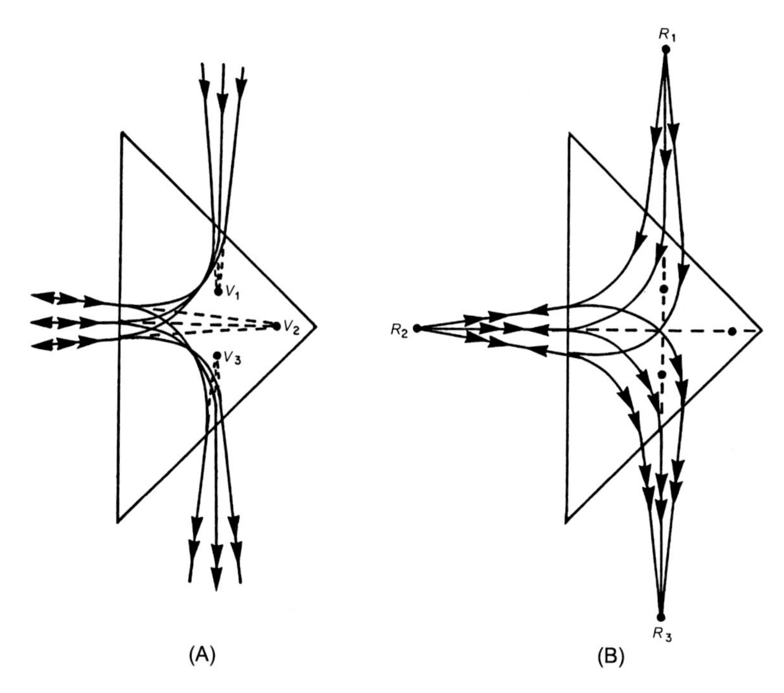

Figure 40 (A) The virtual stigmatic points V_1, V_2, and V_3. (B) The real stigmatic points R_1, R_2, and R_3.

second set of cardinal points corresponds to trajectories which lie in the cylindrical surface whose generators are perpendicular to the first principal section and pass through the mean trajectory $APP'\ A'_1\ A'_2$ (Fig. 41A). This cylindrical surface is called the *second principal section.* Expressions for the positions of the cardinal points have been given by Cotte (1938). In his system of coordinates, the cardinal points lying in object space are measured from the origin P (Fig. 41B) in a direction $\mathbf{T}_1$ tangential to the mean trajectory at P and those in image space from the origin P' and in a direction $\mathbf{T}_2$ tangential to the mean trajectory at P'. These distances are positive if they are parallel to the forward sense of the mean trajectory and negative if antiparallel. The sign convention for ϵ_1 and ϵ_2 is as follows. The angles ϵ_1 and ϵ_2 are positive if $\mathbf{n}_1.\mathbf{T}_1$ and $\mathbf{n}_2.\mathbf{T}_2$, where $\mathbf{n}_1$ and $\mathbf{n}_2$ are the outward normals of the prism faces (Fig. 41B), are also positive. In Fig. 41B therefore, ϵ_1 is negative and ϵ_2 is positive. Cotte's equations are given in Table 2, where f and h are the distances of the focal and principal points in object space and f' and h' are the distances of the focal and principal points in image space.

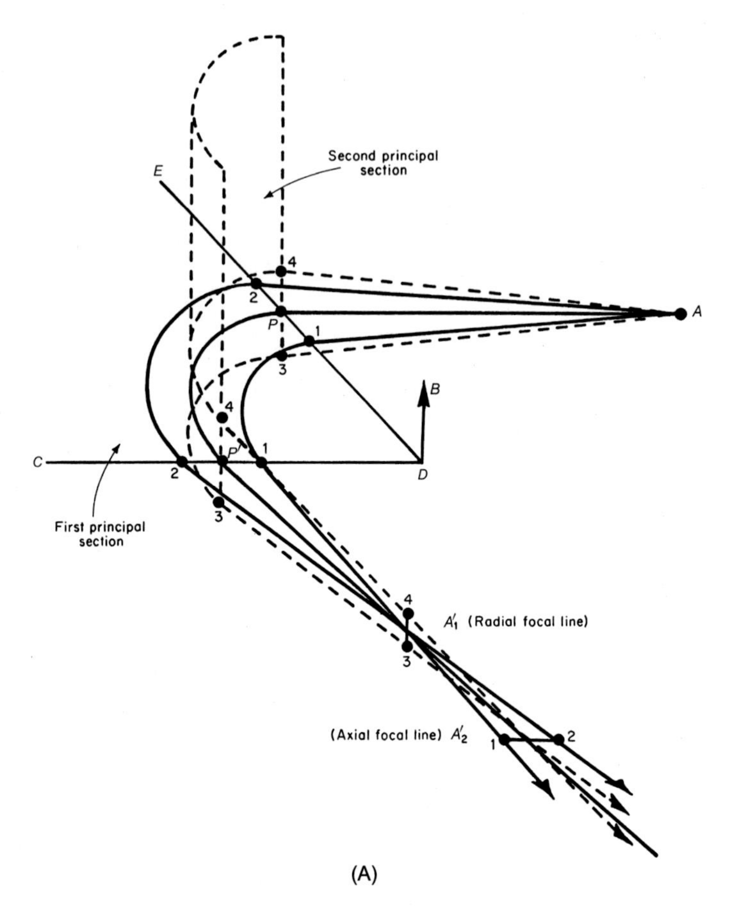

Figure 41 (A) The first and second principal sections of the magnetic prism. (B) A diagram of the first principal section showing the various angles of incidence, etc.

For the first half of the double prism ($\theta = \pi/2$, $\epsilon_1 = -\epsilon$, and $\epsilon_2 = 0$), the cardinal points of the first principal section are given by

$$f_1 = 0; \qquad h_1 = a; \qquad f_1' = a \tan\epsilon; \qquad h_1' = a(\tan\epsilon - 1)$$

and those of the second principal section by

$$f_2 = -a \cot\epsilon; \qquad h_2 = 0; \qquad f_2' = a(\cot\epsilon - \pi/2); \qquad h_2' = -a\pi/2$$

For the second half of the prism ($\theta = \pi/2$, $\epsilon_1 = 0$, $\epsilon_2 = \epsilon$) the cardinal points of the first principal section are given by

$$f_1 = -a \tan\epsilon; \qquad h_1 = -a(\tan\epsilon - 1); \qquad f_1' = 0; \qquad h_1' = -a$$

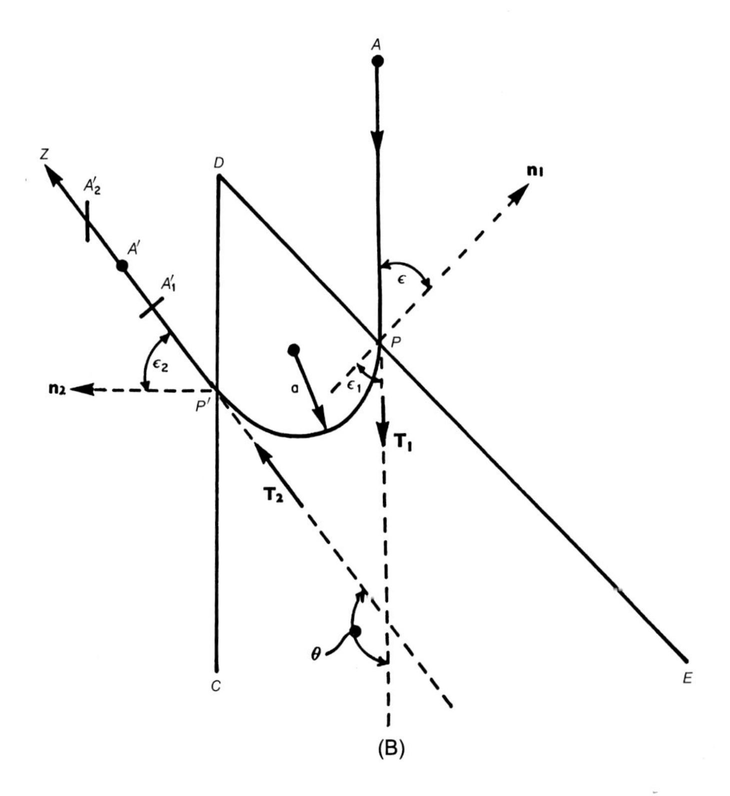

Figure 41 (*continued*)

and for the second principal section by

$$f_2 = -a(\cot\epsilon - \pi/2); \qquad h_2 = a\pi/2; \qquad f_2' = a\cot\epsilon; \qquad h_2' = 0$$

These equations show that the cardinal points of the first principal section in the first and second halves of the prism are symmetric about the mirror axis R_2V_2 (Fig. 42A). A similar symmetry holds for the cardinal points of the second principal section (Fig. 42B). Corresponding cardinal points of the two principal sections do not in general coincide. The radial symmetry of a conical beam of electrons incident on the entrance face of the double prism is therefore not preserved in general as it passes through the system.

For *certain* positions of the object point A, however, the radial and axial focal lines coincide at some point A' (Fig. 41B) and for these positions of A, A' is a stigmatic image point. The conditions which must be met for A_1' and A_2' to coincide can be obtained as follows. Cotte's relations (Table 2) show that the focal distances $P'A_1'$ and $P'A_2'$ (Fig. 41B) are related to a, ϵ_1,

Table 2 Equations given by Cotte (1938) for the distances of the focal and principal points of the magnetic prism

First principal section	Second principal section
$f_1 = -\frac{a\cos(\theta-\epsilon_2)\cos\epsilon_1}{\sin(\theta+\epsilon_1-\epsilon_2)}$	$f_2 = \frac{a(\theta\tan\epsilon_2-1)}{\theta\tan\epsilon_1\tan\epsilon_2-\tan\epsilon_1+\tan\epsilon_2}$
$h_1 = -\frac{a\{\cos(\theta-\epsilon_2)\cos\epsilon_1+\cos\epsilon_1\cos\epsilon_2\}}{\sin(\theta+\epsilon_1-\epsilon_2)}$	$h_2 = \frac{a\theta\tan\epsilon_2}{\theta\tan\epsilon_1\tan\epsilon_2-\tan\epsilon_1+\tan\epsilon_2}$
$f_1' = \frac{a\cos\epsilon_2\cos(\theta+\epsilon_1)}{\sin(\theta+\epsilon_1-\epsilon_2)}$	$f_2' = \frac{a(\theta\tan\epsilon_1+1)}{\theta\tan\epsilon_1\tan\epsilon_2-\tan\epsilon_1+\tan\epsilon_2}$
$h_1' = \frac{a\{\cos(\theta+\epsilon_1)\cos\epsilon_2-\cos\epsilon_1\cos\epsilon_2\}}{\sin(\theta+\epsilon_1-\epsilon_2)}$	$h_2' = \frac{a\theta\tan\epsilon_1}{\theta\tan\epsilon_1\tan\epsilon_2-\tan\epsilon_1+\tan\epsilon_2}$

f and h refer to object space whereas f' and h' refer to image space (see Figs. 42A and B).

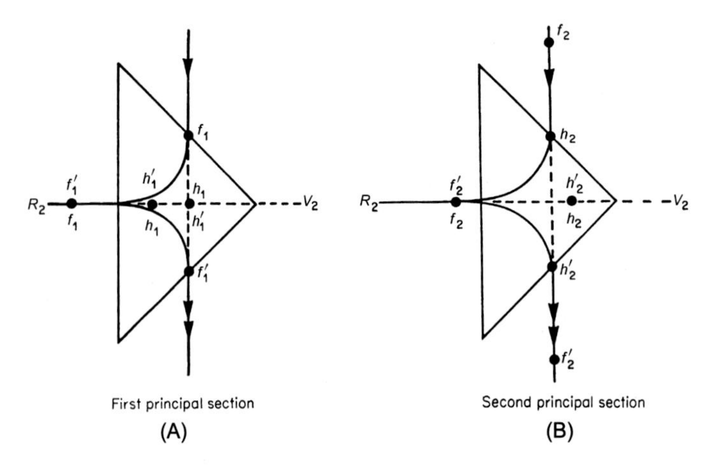

Figure 42 (A) The cardinal points of the first principal section ($\theta < \tan\epsilon < 1$). (B) The cardinal points of the second principal section ($\cot\epsilon > \pi/2$).

ϵ_2, θ, and the object distance PA, by the equations

$$\xi_1'/a = \frac{\frac{(\xi/a)\cos(\theta+\epsilon_1)}{\cos\epsilon_1} - \sin\theta}{(\xi/a)\frac{\sin(\theta+\epsilon_1-\epsilon_2)}{\cos\epsilon_1\cos\epsilon_2} + \frac{\cos(\theta-\epsilon_2)}{\cos\epsilon_2}} \tag{4}$$

$$\xi_2'/a = \frac{\theta - (\xi/a)(1+\theta\tan\epsilon_1)}{(\xi/a)(\tan\epsilon_1 - \tan\epsilon_2 - \theta\tan\epsilon_1\tan\epsilon_2) - (1-\theta\tan\epsilon_2)} \tag{5}$$

where $\xi = PA$, $\xi_1' = P'A_1'$, and $\xi_2' = P'A_2'$. The condition required for coincidence of the radial and axial focal lines is just

$$\xi_1' = \xi_2' = \xi'$$

where $\xi' = P'A'$. The sign convention for ξ and ξ' is the same as that for f, f', h, and h' (see above) and therefore in Fig. 41B ξ is negative and ξ' is positive. For the first half of the double prism $\theta = \pi/2$, $\epsilon_1 = -\epsilon$, and $\epsilon_2 = 0$ with the result that Eqs. (4) and (5) reduce to

$$\frac{\xi_1'}{a} = \tan\epsilon - \frac{1}{(\xi/a)} \tag{6}$$

$$\frac{\xi_2'}{a} = \frac{(\xi/a)}{1 + (\xi/a)\tan\epsilon} - \frac{\pi}{2} \tag{7}$$

Using the condition $\xi_1' = \xi_2' = \xi'$ for the coincidence of the axial and radial focal lines, and eliminating ξ' from Eqs. (6) and (7), we obtain,

$$\left(\tan^2\epsilon + \frac{\pi}{2}\tan\epsilon - 1\right)\left(\frac{\xi}{a}\right)^2 + \frac{\pi}{2}\left(\frac{\xi}{a}\right) - 1 = 0 \tag{8}$$

The solutions for ξ obtained from this equation give the positions of the stigmatic object points as a function of the angle of incidence ϵ. The variation of ξ as a function of $\tan\epsilon$ obtained by solving Eq. (8) is shown in Fig. 43. On eliminating ξ from Eqs. (6) and (7) we obtain

$$\left(\frac{\xi'}{a}\right)^2 + \left(\frac{\pi}{2} - 2\tan\epsilon\right)\left(\frac{\xi'}{a}\right) + 1 - \pi\tan\epsilon = 0 \tag{9}$$

and the solutions of ξ' obtained from this equation give the positions of the stigmatic image points as a function of ϵ. The variation of ξ' as a function of $\tan\epsilon$ obtained from Eq. (9) is also given in Fig. 43. Eqs. (8) and (9) also show that ξ and ξ' have imaginary roots only, for values of $\tan\epsilon$ lying between $-1 - (\pi/4)$ and $+1 - (\pi/4)$, and within this range of values of $\tan\epsilon$ the prism does not possess stigmatic points. In addition, for values of tan greater than $\frac{1}{4}\{(\pi^2 + 16)^{\frac{1}{2}} - \pi\}$ the prism always possesses two pairs of stigmatic points, one pair being real and the other virtual.

So far only the relationship between the positions of the stigmatic points in the upper half of the double prism have been considered. After reflection by the mirror, however, the beam traverses the lower half of the prism (Fig. 44) and it will now be shown that the latter possesses stigmatic points which are symmetric to those in the upper half about the axis R_2V_2 (Fig. 39). The mean trajectory re-enters the double prism with an angle of incidence $\epsilon_1 = 0$ and after suffering a total angular deviation of $\pi/2$, emerges at an angle $\epsilon_2 = \epsilon$. Substitution of these values in Eqs. (4) and (5), followed by application of the stigmatic condition $\xi_1' = \xi_2' = \xi$ and successive

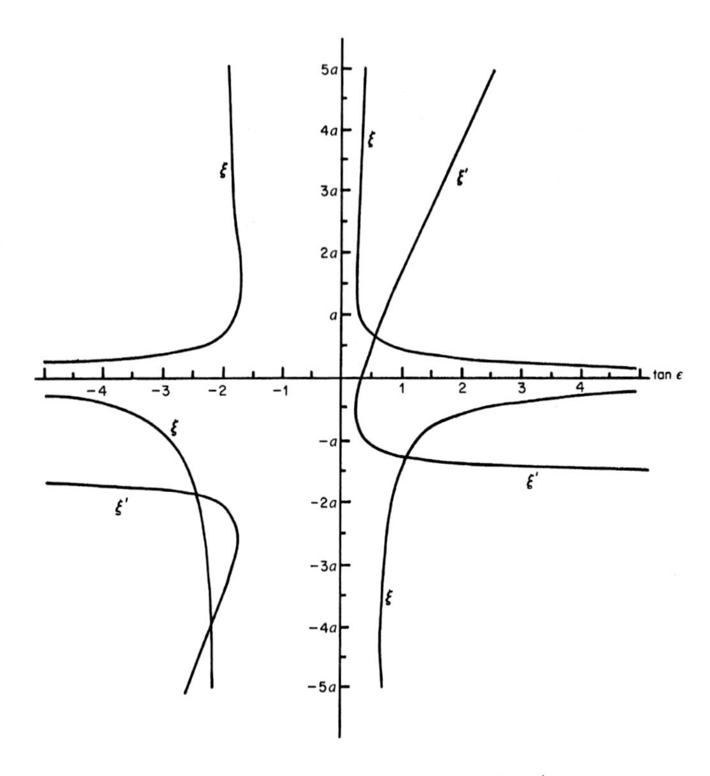

Figure 43 Variation of the image and object distances ξ and ξ' with tan ϵ.

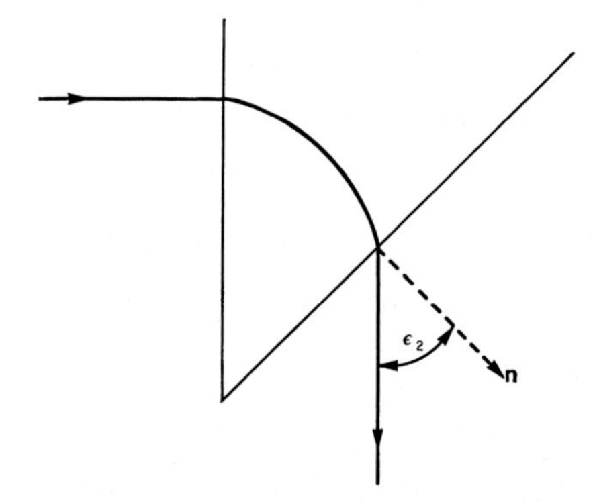

Figure 44 The trajectory of an electron in the second half of the double prism.

elimination of ξ and ξ' yields the equations,

$$\left(\frac{\xi}{a}\right)^2 - \left(\frac{\pi}{2} - 2\tan\epsilon\right)\left(\frac{\xi}{a}\right) + 1 - \pi\,\tan\epsilon = 0 \tag{10}$$

$$\left(\tan^2\epsilon + \frac{\pi}{2}\tan\epsilon - 1\right)\left(\frac{\xi'}{a}\right)^2 - \frac{\pi}{2}\left(\frac{\xi'}{a}\right) - 1 = 0 \tag{11}$$

A comparison of these equations with Eqs. (8) and (9) shows that the only differences are that the object and image points in the first half of the prism correspond to image and object points in the second half, and that ξ and ξ' in Eqs. (10) and (11) are of opposite sign to those in Eqs. (8) and (9). The stigmatic points of the double prism are therefore situated symmetrically about the axis R_2V_2 (Fig. 39) and furthermore the stigmatic image points of the first prism are coincident with the stigmatic object points of the second prism. In practice this coincidence can only be brought about if the apex of the mirror and its center of curvature are at R_2 and V_2 (Fig. 39). If the apex is at R_2 and the center at V_2 the system is dispersive, whereas if the apex is at V_2 and the center at R_2 it can be shown that the device is completely achromatic, and hence the latter configuration is of no interest if the system is intended for use as an energy analyzer.

From the preceding analysis it is clear that if an incident beam has its cross-over situated at R_1 or V_1 (Fig. 39), the radial symmetry of the beam is not destroyed by its passage through the mirror-prism system. In practice the filter lens is placed in a position such that the stigmatic object and image points R_1 and V_1 lie in the back focal plane and image plane respectively of the objective lens (Fig. 45) of an electron microscope. In this case a stigmatic image is produced in the plane I_2 passing through the point V_3 and the cross-over of the exit beam is produced at R_3. It is clear from the symmetry of the stigmatic points R_1, V_1, R_3, and V_3 about the mirror axis that the magnification produced by the mirror-prism system is unity, and this can be proved rigorously from Cotte's equations (Table 2). The filter lens, therefore, simply inverts the positions of the cross-over and image produced by the objective lens.

So far in this discussion it has been assumed that the electrons passing through the system are monoenergetic. The effect of the filter lens on electrons of different energies will now be considered and the dispersive properties of the system examined. If an electron of energy E follows a trajectory PP' (Fig. 46) of radius a, then an electron of energy $E + \Delta E$ will follow a trajectory PP'_1 of radius $a + \Delta a$. It will be assumed that Δa is small so that to the first order $P'P'_1 = \Delta a$. The calculation of the trajectory of an electron of energy $E + \Delta E$ through the mirror-prism system is largely a matter of geometry. The coordinates of the various points P, P', etc., indicated in Fig. 46 are given in Table 3 and these have been obtained by neglecting second order quantities $(\Delta a)^2$ etc., and by assuming

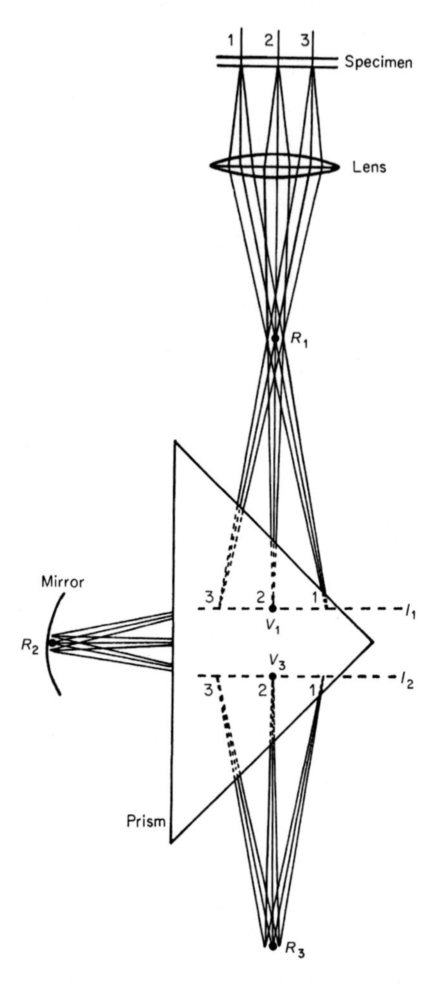

Figure 45 If the image plane of an objective (or intermediate) lens is made coincident with the virtual stigmatic object plane I_1, a real stigmatic and achromatic image is formed in the plane I_2 after reflection by the electrostatic mirror.

$\tan\epsilon > \frac{1}{4}\{(\pi^2 + 16)^{\frac{1}{2}} - \pi\}$. It is also assumed that the signs of the stigmatic object and image distances ξ and ξ' correspond to those given in Fig. 43, these being the values for the first half of the double prism. The positive value of ξ in Fig. 43 is denoted by ζ_1, the negative value by ζ_2; the positive value of ξ' by ζ_1' and the negative by ζ_2'.

The trajectory PP' is given by the equation

$$\{x - (a + \Delta a)\}^2 + (y - a)^2 = (a + \Delta a)^2$$

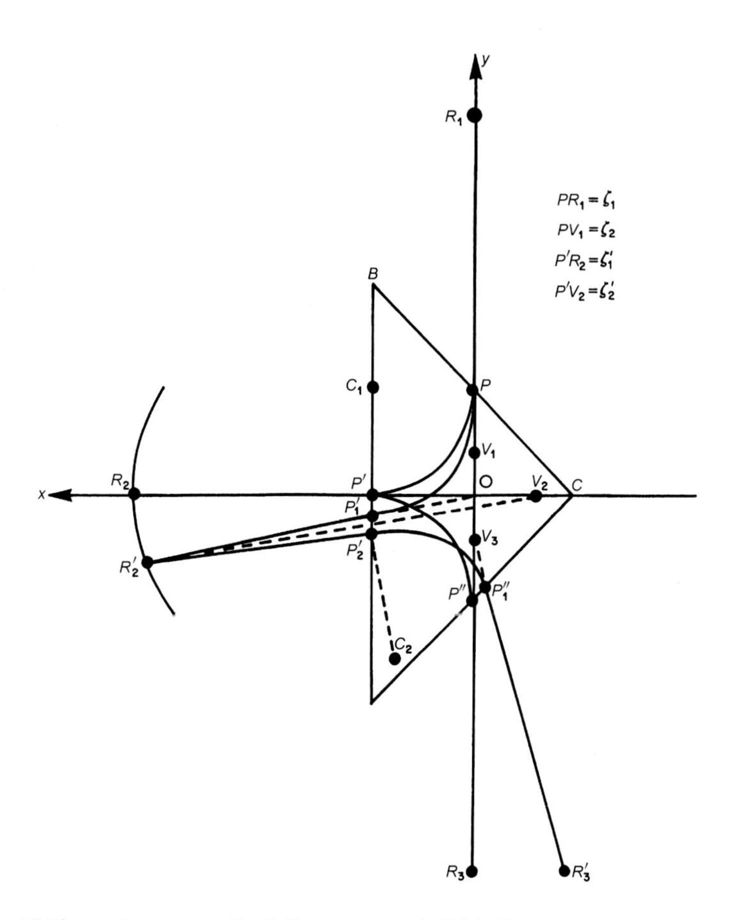

Figure 46 The various points R_1, P, C_1, etc. given in Table 3.

and the equation of the tangent $P_1'R_2'$ is

$$y = -(\Delta a/a)x$$

to the first order of small quantities. Electrons of different energies emerging from the face AB of the prism therefore appear to originate from the point 0, which is an achromatic point of the system. An analysis of the trajectories of electrons of different energies originating from different points of a specimen (Fig. 47) shows that an achromatic (and also stigmatic) image of the specimen is formed on the plane defined by the equation $x = 0$ (Fig. 47), and this plane is therefore an achromatic plane of the system. After reflection by the mirror, electrons of energy $E + \Delta E$ follow the circular path $P_2'P_1''$ (Fig. 46), whose center is at C_2, and emerge from the prism

Table 3 Positions of the various points given in Fig. 46

Point	x	y
P	0	a
P'	a	0
P'_1	a	$-\Delta a$
P'_2	a	$-\frac{\Delta a}{a}\left\{\frac{a(\zeta'_1+\zeta'_2)+2\zeta'_1\zeta'_2}{(\zeta'_2-\zeta'_1)}\right\}$
P''	0	$-a$
P''_1	$2\Delta a\frac{(\zeta'_1+a)}{(\zeta'_2-\zeta'_1)}$	$-a-2\Delta a\frac{(\zeta'_1+a)}{(\zeta'_2-\zeta'_1)}\tan\epsilon$
C_1	a	a
C_2	$a+\Delta a\frac{(\zeta'_1+\zeta'_2+2a)}{(\zeta'_2-\zeta'_1)}$	$-a-\frac{2\Delta a}{a}\zeta'_2\frac{(a+\zeta'_1)}{(\zeta'_2-\zeta'_1)}$
R_1	0	$a-\zeta_1$
R_2	ζ'_1+a	0
R_3	0	ζ_1-a
V_1	0	$a-\zeta_2$
V_2	ζ'_2-a	0
V_3	0	ζ_2-a

at P''_1. The equation of the exit trajectory $P''_1R'_3$ is, to the first order of small quantities,

$$y=\frac{a-(\zeta'_1-\zeta'_2)x}{2\Delta a(\zeta'_1+a)\{(\zeta'_2/a)-\tan\epsilon\}}-a\left\{\frac{1}{(\zeta'_2/a)-\tan\epsilon}+1\right\}$$

Eq. (6) shows however, that under the stigmatic conditions assumed,

$$\frac{1}{(\zeta'_2/a)-\tan\epsilon}=\frac{\zeta_2}{a}$$

and therefore that the equation of the trajectory $P''_1R'_3$ can be written as

$$y=\frac{(\zeta'_1-\zeta'_2)\zeta_2}{2\Delta a(\zeta'_1+a)}x+\zeta_2-a \tag{12}$$

It is clear from this equation that the exit trajectory intersects the y axis of Fig. 46 at the virtual stigmatic object point V_3, and that V_3 is another achromatic point of the system. A general analysis shows that the mirror prism device possesses a second achromatic plane defined by the equation $y=\zeta_2-a$, and it follows that if R_1 and V_1 (Fig. 46) are coincident with the back focal plane and image plane respectively of the objective lens of

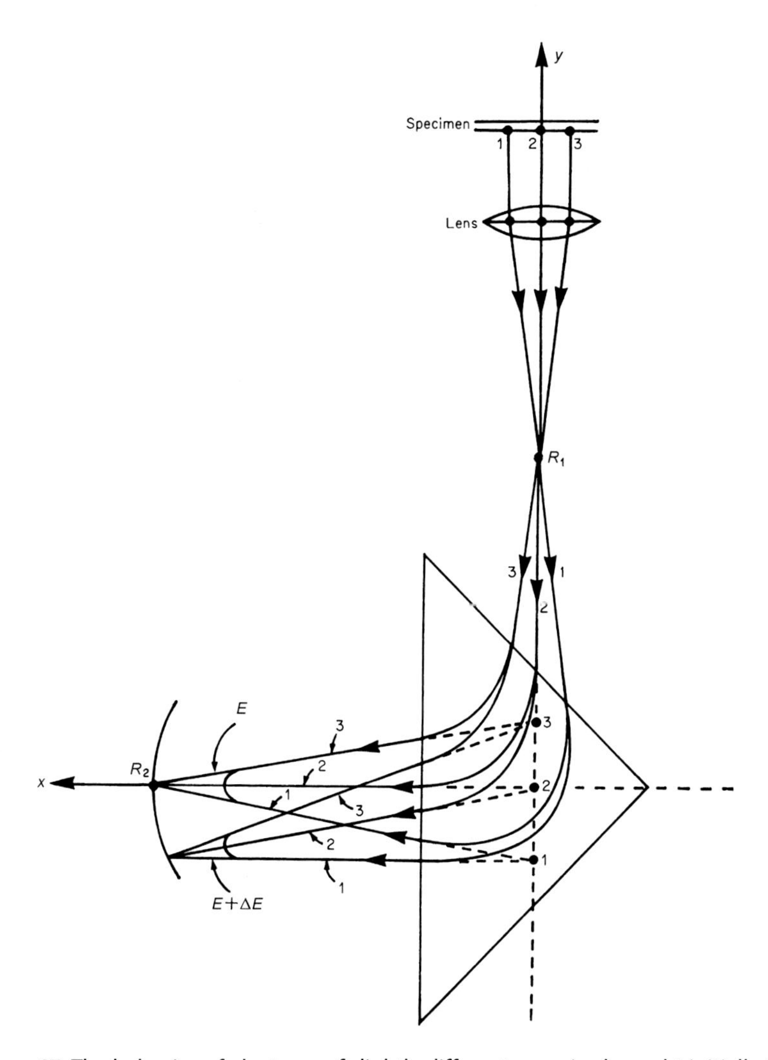

Figure 47 The behavior of electrons of slightly different energies brought initially to a cross-over at the point R_1.

the microscope, then a stigmatic and achromatic image of the specimen is formed at the plane $y = \zeta_2 - a$. The position of the cross-over point on the exit side of the double prism does however depend on the electron energy; those with energy E have their crossover at R_3 (Figs. 46 and 48), whereas those of energy $E + \Delta E$ have their cross-over at R_3'. If the achromatic plane defined by $y = \zeta_2 - a$ is projected on the observation screen of the microscope (Fig. 49) the image so obtained is the same as that which would be

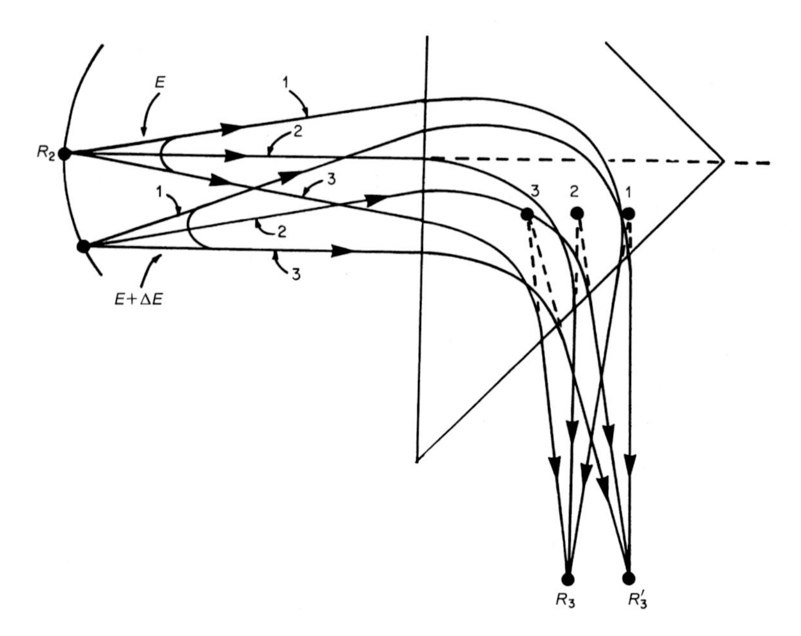

Figure 48 The behavior of electrons of energies E and $E + \Delta E$ in traversing the second halt of the magnetic prism.

produced by a conventional electron microscope. If the plane containing the exit cross-over points of the dispersive system is projected on the final screen, the energy spectrum of the electrons selected by the objective aperture of the microscope will be observed. In this mode of operation however the apparatus does not function as an energy analyzing electron microscope since the observed loss spectrum is produced by electrons which originate over the entire area of irradiation of the specimen. Finally if a selecting aperture is placed at the exit cross-over plane of the filter lens (Fig. 49) and the achromatic plane $y = \zeta_2 - a$ is projected on the final screen, the image so produced is formed by electrons with energies lying between E and $E + \delta E$, where δE depends on the width of the selecting aperture and the dispersion of the system at the exit cross-over plane. An expression for the dispersion can be obtained by determining the point of intersection of the exit trajectory $P_1''R_3'$ (Fig. 46) with the line $y = \zeta_1 - a$. Eq. (12) shows that the x coordinate of R_3' is given by

$$\Delta x = \frac{2\Delta a(\zeta_1 - \zeta_2)(\zeta_1' + a)}{\zeta_2(\zeta_1' - \zeta_2')}$$

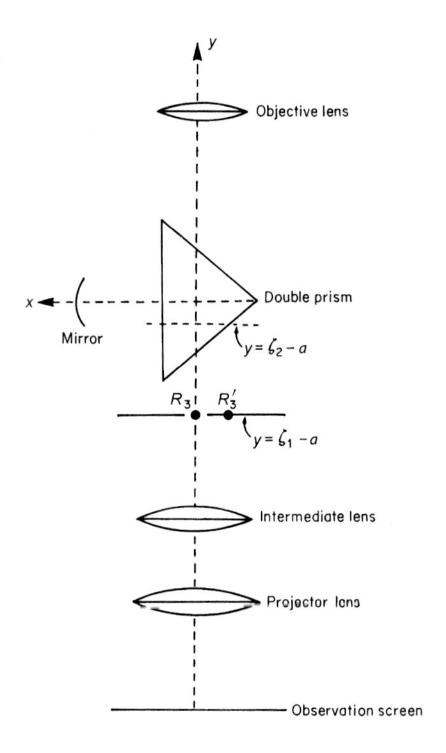

Figure 49 An outline of the energy selecting electron microscope.

If the beam potential of an electron of energy E is V then the radius of curvature a of the trajectory in the prism is proportional to $\sqrt{V}$, with the result that

$$\frac{\Delta a}{a} = \frac{1}{2}\frac{\Delta V}{V}$$

and the dispersion is therefore given by

$$\frac{\Delta x}{\Delta V} = \frac{a}{V}\frac{(\zeta_1 - \zeta_2)(\zeta_1' + a)}{\zeta_2(\zeta_1' - \zeta_2')}$$

The dimensionless quantity $(V/a)|\Delta x/\Delta V|$ is plotted in Fig. 50 as a function of the tangent of the angle ϵ for values of

$$\tan\epsilon > \frac{1}{4}\left\{\left(\pi^2 + 16\right)^{\frac{1}{2}} - \pi\right\}.$$

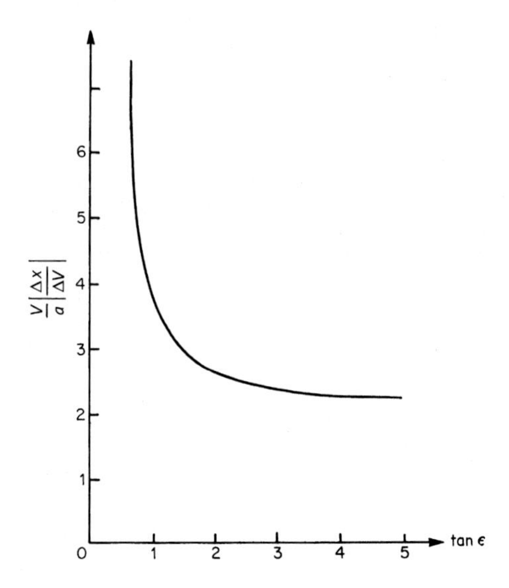

Figure 50 The dimensionless quantity $(V/a)|\Delta x/\Delta V|$ plotted as a function of $\tan\epsilon$. The quantity $|\Delta x/\Delta V|$ is the dispersion of the filter lens.

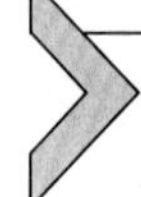

3. ENERGY ANALYZING ELECTRON MICROSCOPES

3.1 Instrumentation

The idea of incorporating an energy analyzer with an electron microscope is not new. The first attempts at combining these instruments to produce a microanalyzer of high spatial resolving power were reported by Hillier (1943) and Marton (1944). Both authors used homogeneous field magnetic analyzers of poor resolving power ($E/\Delta E_r \sim 10^3$), with the result that much of the fine detail of the energy loss spectra was lost. Furthermore the state of electron microscopy at this period was such that only shadow images of the specimen could be obtained, and it was therefore impossible to relate changes in the loss spectra to changes in the substructure of the specimens examined.

The work of Hillier and Marton was followed by that of Kleinn (1954), Leonhard (1954), Marton and Leder (1954), Gauthe (1954), and Watanabe (1954, 1955, 1956). These authors employed the high resolution electrostatic analyzer described in Section 2.1 of this article. In each case the analyzer was placed between the intermediate and projector lenses of the microscope. This position is highly undesirable if full use of the spatial resolution and magnification of modern electron microscopes is to be made. Consider an analyzer employing an entrance slit of width 5 μm situated in the image plane of the intermediate lens of a microscope producing an overall magnification of 20,000× at the final image plane. The objective lens forms an image in the object plane of the intermediate lens and this first stage of magnification is typically 25×. The intermediate lens further magnifies the image by a factor of about 6.4 so that at the analyzer entrance aperture the magnification is 160×. The apparent width of the entrance slit, which for the moment will be taken as the spatial resolution, is therefore (5/160) μm or 310 Å. If, however, use is made of the magnification of the projector lens and the analyzer is mounted in the final image plane where the magnification is 20,000×, the apparent width of the slit is 2.5 Å. An analyzer placed in the latter position therefore takes full advantage of the spatial resolution (~3 Å) afforded by commercial electron microscopes, and selects, in effect, a line in the final image from which the energy spectrum is recorded. One further advantage associated with placing the analyzer in this position is that by simply cutting a narrow slot in the final screen of the microscope, simultaneous observation of both image and spectrum is achieved. It is only in recent years that sufficiently sophisticated instruments have been developed to allow realization of Hillier's and Marton's original aim of high resolution microanalysis. The rest of this section is devoted to a discussion of energy analyzing electron microscopes which have been developed in the Cavendish Laboratory.

The first system to be described here consists of a cylindrical electrostatic analyzer (Section 2.1) mounted below the final image screen of a Siemen's Elmiskop I microscope operating at 100 kV (Cundy, Metherell, & Whelan, 1966; Metherell, 1965; Metherell, Cundy, & Whelan, 1965). A schematic outline of the instrument is shown in Fig. 51. The analyzer section extends from the base of the camera chamber C_2 of the microscope to the floor. This extension of the microscope column means that the control console upon which C_2 usually rests must be removed and mounted on a separate trolley. The entrance slit unit S of the analyzer is suspended from a trolley T which allows translation and rotation of the slit assembly for alignment purposes. The trolley is housed in the camera chamber C_2 which means that the image cannot be recorded in the usual manner on photographic plates stored in C_2. For this reason a projector tube camera C_1, mounted on the viewing port chamber, is employed to record the images on 35 mm film. The camera unit in C_1 can be swung into a horizontal position for recording purposes and into a vertical position to allow visual observation of the image on the final screen. The length and width of the aperturing slit of the analyzer can be altered by using controls mounted

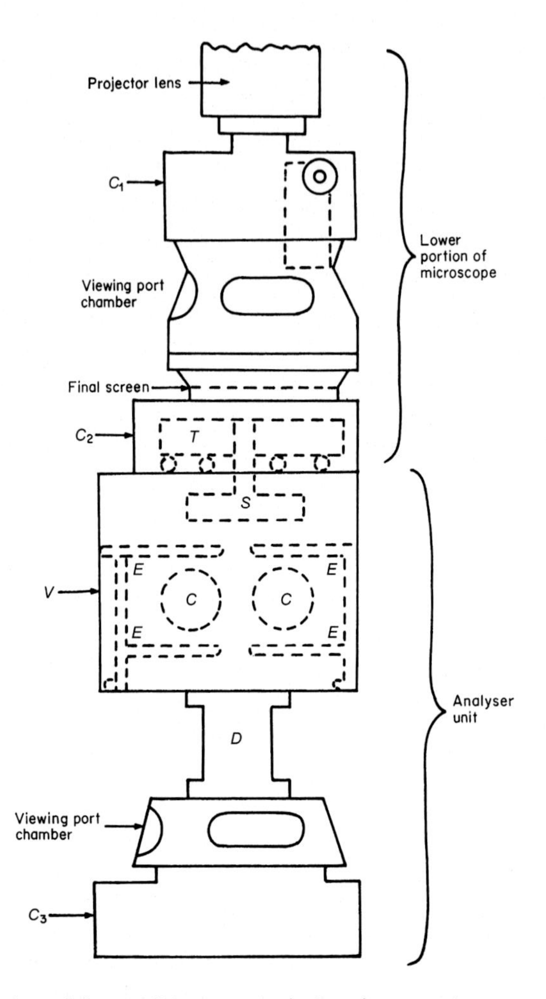

Figure 51 An outline of the 100 kV energy analyzing electron microscope which utilizes the cylindrical electrostatic analyzer described in Section 2.1.

out of vacuum and coupled to the slit assembly by sliding connector rods and universal couplings. The electrode system of the analyzer (C and E) is housed in a vacuum chamber V which is attached to a tube D. The function of this tube is to increase the projection distance and hence obtain increased dispersion in the spectrum. The dispersion tube is in turn mounted on a viewing port chamber which allows visual observation of the spectra, which are recorded photographically on plates stored in the camera chamber C_3.

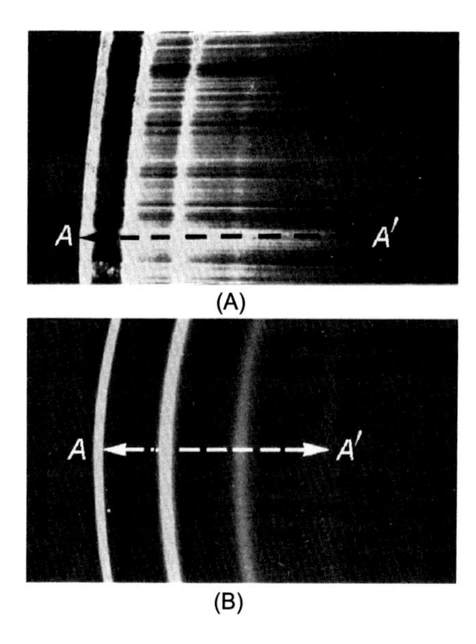

Figure 52 (A) An example of the spectrum obtained when a spectrographic slit is used as e entrance aperture of the analyzer. (B) An example of the spectrum obtained when the edge pieces of the slit mechanism have been prepared in the manner discussed in the text. The lines AA' are the energy loss axes of the spectra.

One practical point worthy of note here is the method used to prepare the edges of the analyzer entrance aperture. Slits used in optical spectrometers are usually unsuitable for electron energy analyzers, since the slit widths employed in the latter are of the order of a few microns and small scale roughness of the edges can lead to undesirable streaking effects in the recorded spectra (Fig. 52). The following slit edge preparation technique, suggested by F. Fujimoto (private communication), has been found to be highly successful in practice (Cundy, 1968; Considine, 1970). Each edge piece of a spectrometer slit mechanism is milled down so that the cross-section has the dimensions indicated in Fig. 53. Fine glass fibers of diameter ~0.2 mm are glued to the blunted edges using a suitable adhesive such, as "Araldite", care being taken to ensure that no adhesive flows on to the outside edges of the glass fibers. After the adhesive has dried the slit edges must be thoroughly cleaned with suitable solvents and then coated with a layer of evaporated gold about 1000 Å thick. To provide a tenacious layer of gold the evaporation must be carried out as slowly as possible.

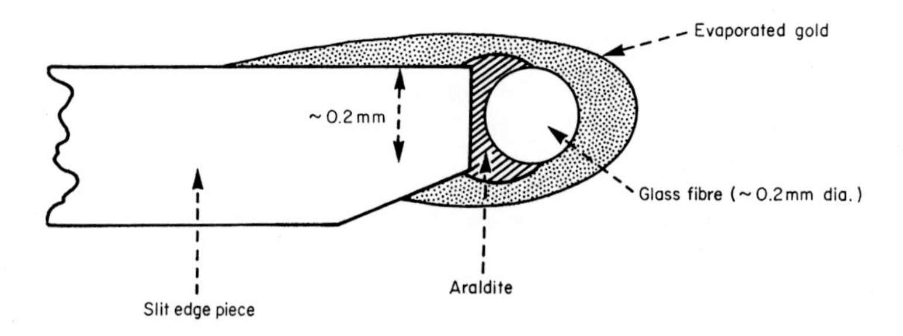

Figure 53 A cross-section of a slit edge piece modified for electron optical use.

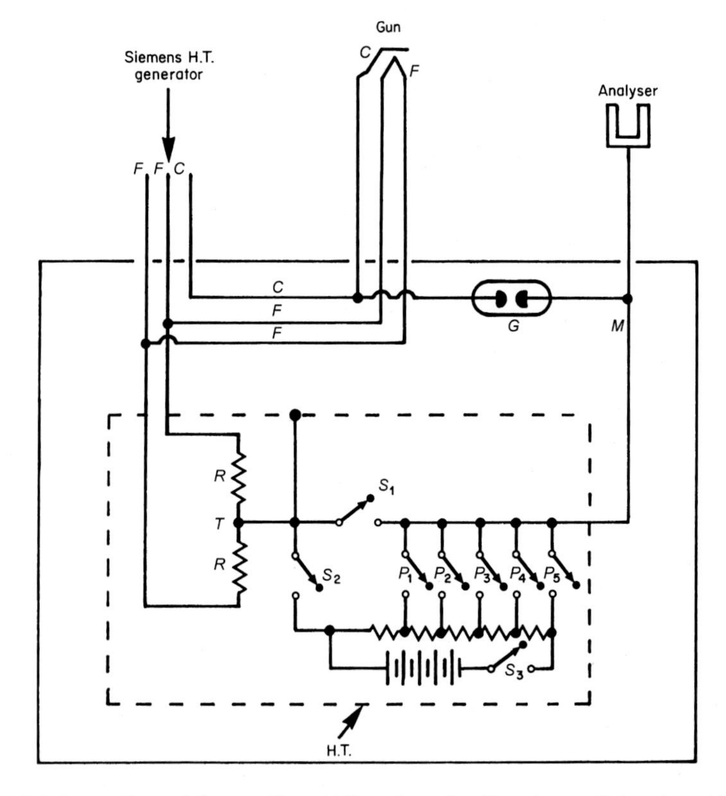

Figure 54 An outline of the auxiliary H.T. tank and calibration unit (enclosed by the broken lines). The switches *S* and *P* are activated by phototransistors triggered by light beams, thus avoiding electrical insulation problems.

The H.T. cable, which normally carries the filament and cathode lines directly to the gun of the microscope, is fed into an auxiliary H.T. tank which is illustrated schematically in Fig. 54. This tank acts primarily as a junction box so that the H.T. applied to the electron gun can also be applied to the center electrodes of the analyzer. The two lines F (Fig. 54) carry the filament heating current and since the beam potential is the voltage difference between the tip of the filament and earth, it is desirable that the analyzer should be biased at the same voltage; the reason being that any small voltage fluctuations produced by the H.T. generator during the time required to record an energy loss spectrum affect both the beam potential and the analyzer bias equally. If the analyzer is biased in this manner no spatial displacement of the loss spectrum occurs at the recording plane, to the first order of small quantities, with the result that these voltage fluctuations do not affect the energy resolution of the instrument. The lead to the analyzer is therefore taken from the center tap T of two equal resistances R (Fig. 54). The resistance of the electron gun filament line is $\sim$10 Ω and to prevent current drain through the resistances R, the latter are made to have values $\sim$50 kΩ.

To obtain accurate calibration of the energy scale of the loss spectra, a potentiometer chain is placed between the center tap T and the output line to the center electrodes of the analyzer. At first sight it would seem that the energy loss calibration requires biasing the filament of the gun *positive* by known increments of voltage. To the first order of small quantities however, exactly the same effect is achieved by biasing the analyzer *negative* by the same increments of voltage. This latter procedure is preferable because the beam brightness is controlled by developing a suitable potential difference between the filament and cathode. As this potential difference is $\sim$ several hundred volts and calibration requires simulation of energy losses up to $\sim$100 V, the former procedure affects the intensity of the incident beam as calibration proceeds. For normal operation the switch S_1 is closed and the switches S_2, S_3, P_1, P_2 etc. are opened. For calibration S_1 is opened, S_2 and S_3 closed, and by successively closing and reopening the switches P_1, P_2 etc. known increments of voltage are added to the H.T. bias to simulate the required energy losses. A safety device worthy of note here is the spark gap G which connects the cathode line to the analyzer lead (Fig. 54). The Siemens H.T. generator is fitted with an electronic device which switches off the H.T. to the gun if the latter sparks over. The safety device is insensitive to appreciable voltage changes of the filament and only acts if the voltage on the cathode line falls below a predetermined level. If the spark gap G (Fig. 54), which fires when a voltage greater than about 800 V is applied across it, is not present in the circuit, then damage to various electronic components in the Siemens' generating unit will occur.

The second system (Considine, 1970; Considine & Smith, 1968), which will now be described, is a cylindrical magnetic analyzer unit attached to the Cavendish 750 kV electron microscope (Smith, Considine, & Cosslett, 1966). A discussion of the energy resolution and dispersion of this system can be found in Section 2.2. The main factors which influenced the design of this energy analyzing attachment are:

(a) the very small dispersion obtained with the magnetic system (~5 μm/V at 100 kV);

(b) the restricted space between the base of the camera chamber and floor of the Cavendish high voltage microscope. Employment of photographic methods for recording loss spectra requires the incorporation of an alignment coil and auxiliary lens with the analyzer unit. If the spatial resolution of photographic emulsion is taken as 100 μm and the dispersion as 5 μm/V at a beam potential of 100 kV, then to achieve an energy resolution of 1 eV an auxiliary lens of magnification 20× is required. The size of a lens producing a magnification of this order, with incident beam potentials up to 750 kV, is such as to prohibit the use of photographic techniques with the Cavendish microscope. This problem was overcome by using electronic methods for recording the loss spectra.

A schematic outline of the unit designed by Considine and Smith (1968) is given in Fig. 55. The incoming electron beam passes through the entrance aperture and is dispersed by the analyzer into its various energy loss components. One of the loss components is deflected in the x direction (Fig. 21), by a D.C. shift coil, on to an energy selecting slit situated immediately above the scintillator of a photomultiplier unit. The loss spectrum is scanned across the selecting slit by means of a second coil, the output of the photomultiplier is amplified and the resulting signal fed into the y amplifier of either an oscilloscope or an xy recorder. The time scale or x axis of the recording is the energy loss axis of the spectrum and the success of this method relies on the linearity of the dispersion as a function of energy loss. With the magnetic system this means that fractional losses up to $\Delta E/E \approx 3 \times 10^{-3}$ (Fig. 36) can be recorded by this method. Alternatively, the scan coil could be dispensed with and instead the excitation of the pole pieces of the analyzer could be slowly reduced. In this case electrons of different energies entering the selecting slit would follow identical trajectories and any non-linearity of the dispersion would have no effect on the energy loss axis of the recorded spectrum. This method can, however, lead to undesirable hysteresis effects in the analyzer pole pieces and since most

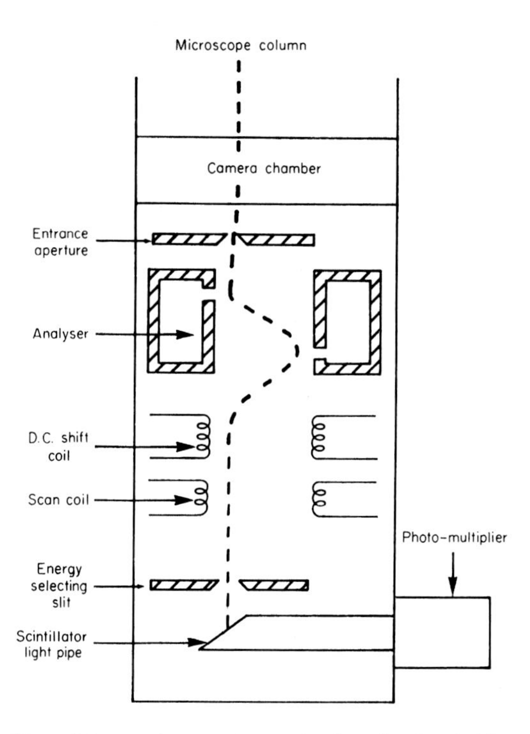

Figure 55 An outline of the analyzer unit attached to the Cambridge 750 kV electron microscope.

of the losses of interest have values $\Delta E/E \leqslant 10^{-3}$, the former method of electronic recording is preferable.

The entrance aperture of the analyzer is circular with diameter ~30 μm. With electronic recording no purpose is served by making this aperture in the form of a slit; the reason being that the photomultiplier unit produces a signal which is proportional to the total intensity incident on the phosphor and if a slit is employed the detector is insensitive to intensity variations along the length of the slit. This puts the system at a serious disadvantage if it is desired to perform microanalysis experiments of the type described in Section 3.2. These experiments can only be carried out with ease if the loss spectra from points lying on a line selected in the image can be recorded simultaneously, as is the case if photographic methods are employed. To obtain the equivalent information when electronic detection is used involves displacing the image repeatedly by small amounts in a given direction and recording the spectrum obtained at each displacement, a process which is at best tedious.

The system can, however, be readily converted into a high voltage energy selecting electron microscope. To achieve this, the scan coil of the unit illustrated in Fig. 55 is dispensed with and the D.C. shift coil is used to deflect a given loss component of the electron beam on to the selecting slit. A set of xy deflecting coils placed between the objective and intermediate lenses of the microscope allows the final image to be scanned in a raster across the entrance aperture of the analyzer. The output of the photomultiplier is used to modulate the brightness of an oscilloscope whose raster is in synchronism with the scan of the microscope image. These modifications, if carried out, would result in the display of an energy selected image on the cathode ray tube of a T.V. display unit.

The two coils of the analyzer each consist of 500 turns of copper wire encapsulated, in vacuum tight aluminum cans. These cans have small flexible pipes which communicate through a vacuum port to the external atmosphere. This alleviates heating, contamination, and vacuum pumping problems which would result if the coils, were run in vacuum. The D.C. shift coil, consisting of 750 turns, and the scan coil, consisting of 375 turns are both wound on the *same* pair of pole pieces, the latter being square in cross-section. The use of square pole pieces, rather than circular, eliminates to a first order approximation astigmatism in the x-direction (Fig. 21). This design does not, of course, eliminate astigmatism in the z-direction, but a focusing action in this direction is not a disadvantage since it is perpendicular to the direction of dispersion. The energy selecting aperture is in the form of a slit, rather than a circular aperture, to alleviate the problem of alignment. It is important that the width w of this slit should be nearly the same as the image width δx_1 of the entrance aperture of the analyzer (Fig. 33). If w is greater than δx_1 then the energy resolution is determined by w rather than by the electron optical properties of the analyzer, and also there is little point in making w less than δx_1 since no increase in the energy resolution occurs, and the only result is a reduction in the intensity arriving at the scintillator of the photomultiplier unit. When electronic methods are used to record the loss spectrum, it is desirable to use values of the excitation parameter k (Section 2.2) close to $k_{\min}$ (Fig. 35) so that only a weak excitation of the D.C. shift coil is needed to deflect the beam on to the selecting slit (Fig. 38). The variation of the image width δx_1 with excitation parameter k (Fig. 35) shows that at $k = k_{\min}$ the slit width which must be employed is ~2 μm. This width is difficult to achieve in practice and a more reasonable value for the lower limit of the slit width is ~4 μm. A value of $\delta x_1 = 4$ μm is obtained when $k = 0.52$ amp. turns $(\text{volt})^{-\frac{1}{2}}$ and this represents the minimum value of k that can be used with the analyzer when electronic methods are used to record the energy spectrum.

Energy loss calibration is carried out by introducing known changes in the accelerating potential of the gun of the electron microscope. The H.T. generator of the Cavendish 750 kV microscope is air insulated and this allows easy access to the electron gun and injector electronics. Insertion of auxiliary equipment inside the torus dome of the accelerator unit allows the beam potential to be changed by known increments and hence calibration of loss spectra is readily achieved.

3.2 Applications

An energy analyzing electron microscope can be operated to record either the loss spectrum of electrons forming a selected area diffraction pattern (Figs. 56A and B) or the spectrum of electrons which contribute to the image of a specimen (Figs. 58 and 59). The diffraction pattern of a specimen of aluminum at a beam potential of 100 kV is given in Fig. 56A and the loss spectrum of electrons arriving along the line SS' in this pattern is shown in Fig. 56B. The lines S_0S_0', S_1S_1', and S_2S_2' indicated in Fig. 56B are the zero loss, first plasmon loss (15 eV) and second plasmon loss (30 eV) lines, respectively, and the familiar parabolic dependence of the plasmon energy loss on the angle of scatter is clearly evident in this spectrum. It is noticeable that the loss lines S_0S_0' are curved (see also Figs. 58B and 59B). This curvature is due partly to end caps fitted to the inner electrodes of the analyzer and partly due to the fact that the trajectories of electrons entering the slit at different points along its length make different angles with the optic axis of the system (Fig. 57). The analyzer is only sensitive to the component of the electron momentum perpendicular to the plane containing the axes of the cylindrical electrodes. Electrons of the same energy arriving at points 1 and 3 (Fig. 57) on the entrance slit have normal components of momenta less than that of an electron arriving at point 2, with the result that the analyzer registers an apparent energy loss for electrons 1 and 3 and this leads to the curvature of the loss lines shown in Figs. 56B and 59B. For an energy *selecting* electron microscope utilizing the Möllenstedt analyzer (Section 4.1) it is important that this curvature be eliminated, since it leads to a first order aberration of the filtered image. For the energy *analyzing* electron microscope however, this curvature is unimportant and needs no correction; the reason being that a loss spectrum is always recorded together with a calibration plate and any distortions or aberrations produced by the system affect both equally.

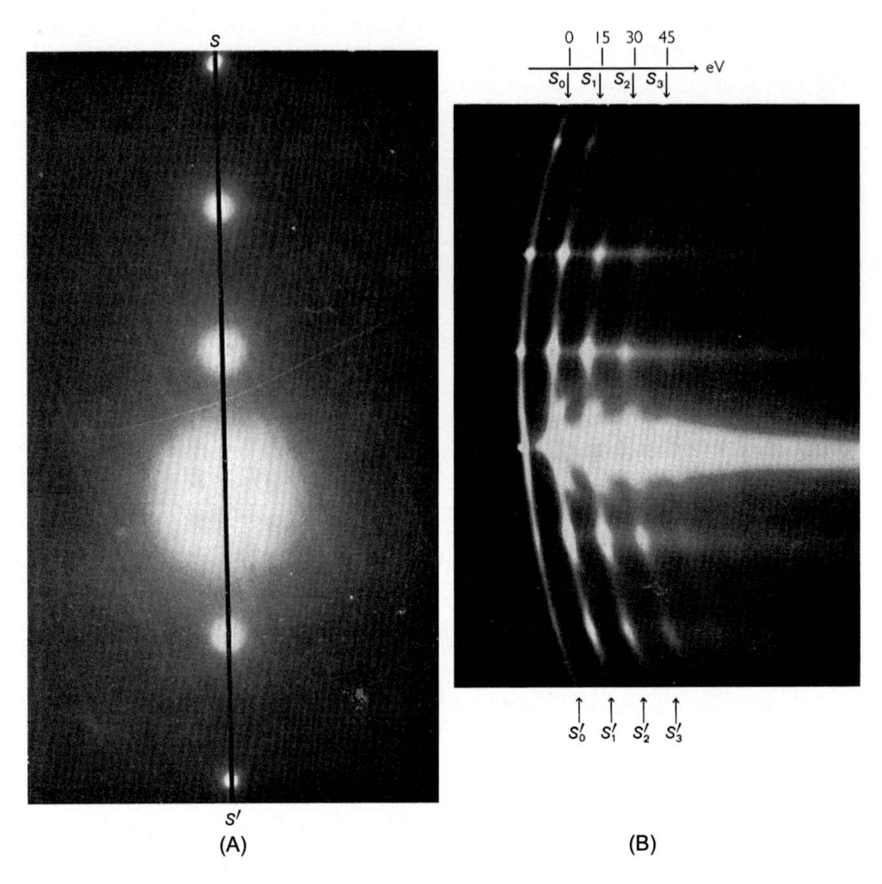

Figure 56 (A) A diffraction pattern of an aluminum specimen and (B) the energy loss spectrum of electrons arriving at the line indicated in the diffraction pattern.

Most of the experiments involving the use of the energy analyzing electron microscope have relied on measurements of the loss spectra of electrons contributing to images rather than diffraction patterns. The reason is that a selected area diffraction pattern contains information averaged out over an area of specimen of the order of several microns across, whereas with an image the spatial resolution, as limited by the width of the analyzer entrance slit, is of the order of several Ångstroms. Bright field thickness fringes obtained with a specimen of aluminum is shown in Fig. 58A. The horizontal line drawn on this micrograph indicates the position of the analyzer entrance slit and the loss spectrum of the electrons contributing to this line in the image is shown in Fig. 58B. In this figure, Z, P_1, and P_2 are the zero loss, first and second plasmon loss lines respectively. In Fig. 58C

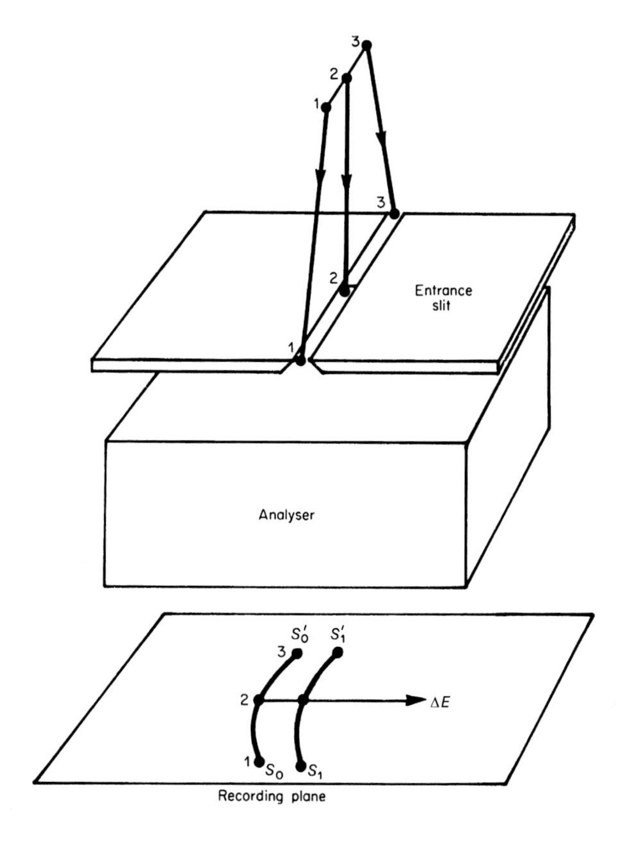

Figure 57 Diagram illustrating one of the reasons why the energy loss lines appear curved (see text for discussion).

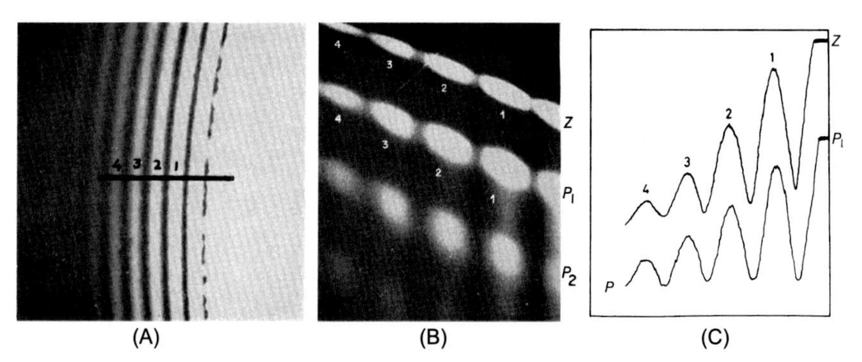

Figure 58 (A) An electron micrograph of a wedge shaped specimen of aluminum. (B) The energy loss spectrum of electrons contributing to the intensity variation along the line indicated in the micrograph. Z, P_1, and P_2 are the zero loss, first and second plasmon losses respectively. (C) Microdensitometer traces taken from the lines Z and P_1 of the spectrum (after Spalding, 1970).

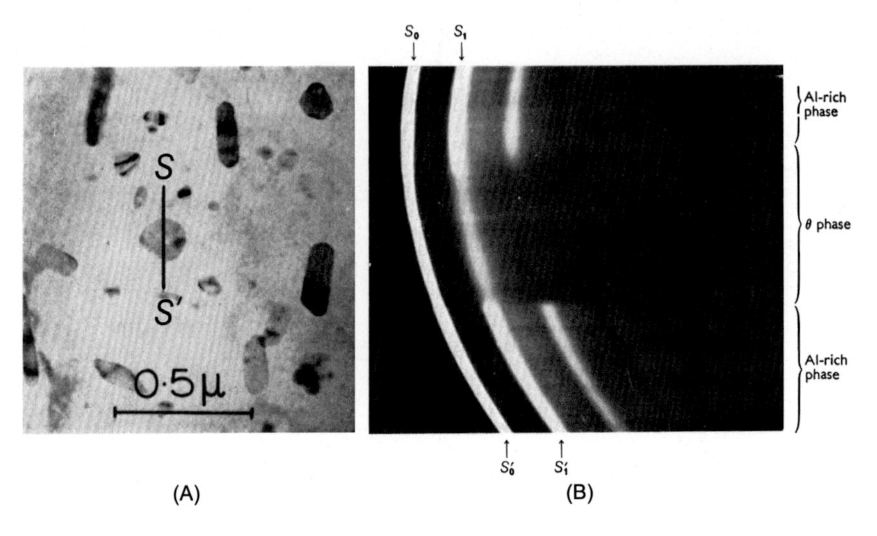

Figure 59 (A) Micrograph of θ phase precipitates in solid solution Al-4% Cu alloy. (B) The loss spectrum of electrons contributing to the intensity along the line indicated in the micrograph (after Cundy, Metherell, & Whelan, 1968, courtesy of *The Philosophical Magazine*).

are shown microdensitometer traces of the zero loss and first plasmon loss components of the spectrum of Fig. 58B, and it is clear from both the loss spectrum itself and the intensity traces of Fig. 58C that the plasmon loss electrons preserve the contrast observed in the image. This preservation of contrast by the inelastically scattered electrons, first-observed by Kamiya and Uyeda (1961), has been the subject of intensive experimental studies involving the use of both energy analyzing (Cundy, Howie, & Valdre, 1969; Cundy, Metherell, & Whelan, 1966, 1967a) and energy selecting electron microscopes (see Section 4.2 for references). The preservation of image contrast by the inelastically scattered electrons allows details appearing in the image to be related to details appearing in the loss spectrum and means that the position of the analyzer slit relative to the image need not be known to any high degree of accuracy.

An example of the loss spectrum obtained in an alloy system in which the composition of the specimen varies across the field of view is given in Fig. 59B. An electron micrograph of θ phase precipitates of composition $CuAl_2$ imbedded in a matrix of nearly pure aluminum is shown in Fig. 59A and the line SS' marked on this micrograph indicates the analyzer entrance slit position. The lines S_0S_0' and S_1S_1' in Fig. 59B indicate the zero loss and first plasmon loss lines of the spectrum corresponding to the line SS' in

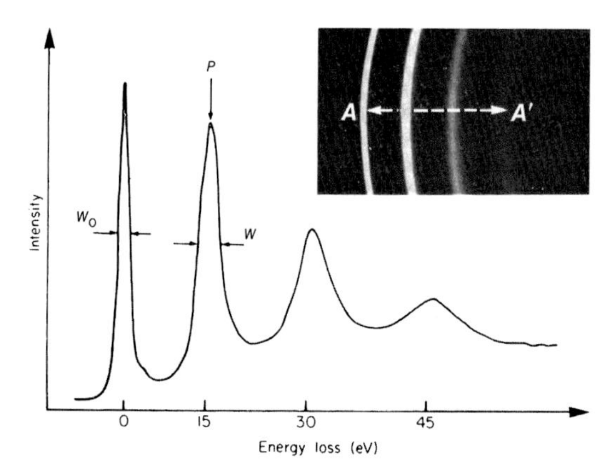

Figure 60 Energy loss spectrum of aluminum. The line *AA′* is the loss axis and *P* is the mean loss (after Spalding & Metherell, 1968, courtesy of *The Philosophical Magazine*).

Fig. 59A and a marked difference between the loss spectrum of the θ phase precipitate and the matrix material is clearly visible. The rest of this section is devoted to a discussion of the principles involved in the microanalysis of binary alloy systems.

The binary alloys most amenable to microanalysis studies are those with constituents which possess sharply defined and well separated energy losses. One of the best examples is the Al–Mg system which has been studied in some detail by Spalding and Metherell (1968). For aluminum the mean plasmon loss, corresponding to the peak P of Fig. 60 is 15.3 ± 0.1 eV, and for magnesium it is 10.4 ± 0.1 eV. The value of the ratio of the loss half-widths w/w_0 is 1.7 for both elements, and since w_0 is typically $\sim$2 eV at 100 kV this gives a halfwidth $w \sim 3.5$ eV. The loss lines of these elements are therefore both well separated and sharply defined. The variation of the mean plasmon loss with alloy composition is given in Fig. 61 and the broken vertical lines of this figure indicate the various phase boundaries of the system (see, for example, Hansen, 1958). The main point of interest here is that for the solid solution a, γ, and δ phases the mean plasmon loss varies linearly with composition, within the limits of experimental error. According to the simple free electron theory of plasmon excitation (see, for example, Raimes, 1967) the energy loss E_p is given by

$$E_p = \hbar\left(4\pi ne^2/m\right)^{\frac{1}{2}}$$

where n is the number of free electrons per unit volume and the other symbols have their usual meaning. The variation of E_p with alloy compo-

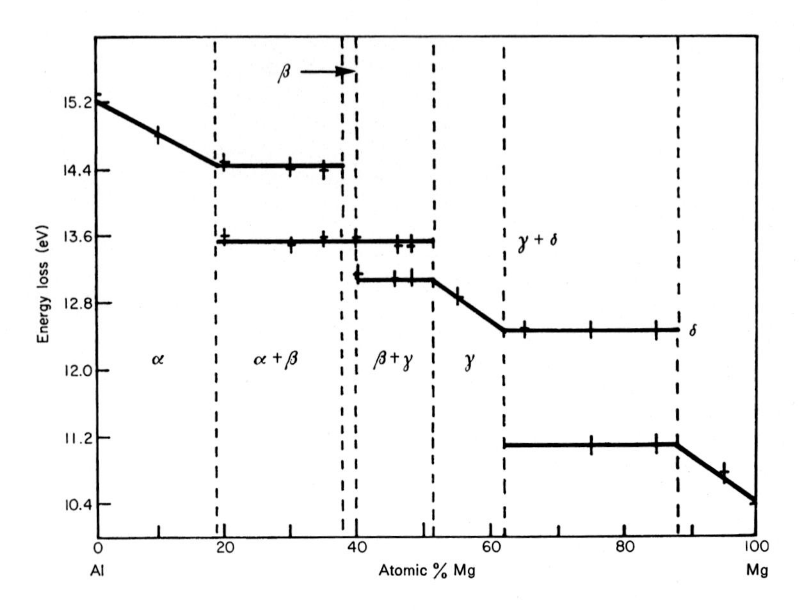

Figure 61 Variation of the mean loss (*P*, Fig. 60) with composition in the Al–Mg system (after Spalding & Metherell, 1968, courtesy of *The Philosophical Magazine*).

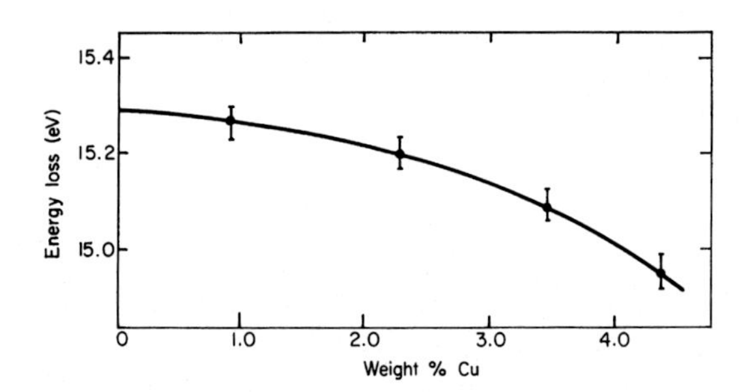

Figure 62 Variation of the mean loss (*P*, Fig. 60) with composition in the Al-rich solid solution phase of the Al–Cu system (after Spalding et al., 1969a, courtesy of *The Philosophical Magazine*).

sition is therefore expected since the addition of magnesium, with two free electrons per atom, to aluminum, with three free electrons per atom, simply dilutes the free electron concentration in the solid solution alloy. Similar measurements of the mean loss in the Al-rich phase of solid solution Al–Cu alloys have been reported by Spalding, Edington, & Villagrana (1969a), and their results are shown in Fig. 62. The non-linear dependence of the loss on copper concentration is due to band structure effects (see for example

Raether, 1965) which are not taken into account in the simple theory of plasma oscillations.

The loss measurements discussed above were made on alloy samples of known composition. It is obvious that if the dependence of energy loss on composition is known then a measurement of the loss can be used to determine the composition of an unknown specimen, and furthermore, if advantage is taken of the spatial resolution afforded by an energy analyzing electron microscope, it must be possible to perform a microanalysis of an inhomogeneous alloy. Other factors unconnected with the alloy composition which could influence the energy loss, such as the effect of vacancy concentration (Cundy, 1968; Cundy, Metherell, & Whelan, 1968) or the effect of elastic constraints in an inhomogeneous alloy (Cook & Howie, 1969), have been considered but are thought to be unimportant in any cases of practical interest that have arisen so far.

Before discussing any specific examples of microanalysis experiments it is pertinent to consider the factors limiting the spatial resolution available with this technique. The spatial resolution as limited by the analyzer entrance slit can be reduced to a value below that of the microscope simply by increasing the magnification of the image sufficiently. A far more important factor limiting the spatial resolution involves the physics of the scattering process. A plasma oscillation in a metal is simply a quantized compression wave of the free electron gas. The plasmon energy is determined by the conduction electron density averaged over some domain in the crystal, the size Δx of this domain being the extent of the plasmon wave packet initially excited by the incident electron. For an inhomogeneous alloy, Δx is therefore the spatial resolution of the microanalysis.

An approximate value for Δx can be obtained from the uncertainty relation

$$\Delta x \Delta q \simeq 1$$

where Δq is the spectral width of the plasmon wave packet in k space. The probability $P(q)$ that the incident electron excites a plasmon into a plane wave state of wave vector q, where q is restricted to have values between $q_{\min}$ and q_c, is proportional to $1/q^2$ (Ferrell, 1956). Defining the width Δq of the probability distribution function as the average value of q, that is by the relation

$$\Delta q = \int_{q_{\min}}^{q_c} \frac{q\,dq}{q^2} \bigg/ \int_{q_{\min}}^{q_c} \frac{dq}{q^2}$$

we obtain on integration

$$\Delta q = \left(\frac{q_c q_{\min}}{q_c - q_{\min}}\right) \ln\left(\frac{q_c}{q_{\min}}\right)$$

The values of $q_{\min}$ and q_c for aluminum are 1.9×10^{-3} Å^{-1} and 2.2×10^{-1} Å^{-1} respectively, giving $\Delta q \approx 10^{-2}$ Å^{-1} and hence that $\Delta x \approx 100$ Å. The spatial resolution as limited by the physics of the inelastic scattering process is therefore ~100 Å and this is considerably larger than that limited by the width of the analyzer entrance aperture.

The principles outlined above have been applied to a study of segregation and initial stages of precipitation in Al-7% Mg alloy (Cundy, Metherell, Whelan, Unwin, & Nicholson, 1968); a study of the dislocation loop growth mechanism in Al-3% Mg alloy (Spalding et al., 1969a); and a study of the copper distribution in lamellar Al–$CuAl_2$ eutectics (Spalding, Villagrana, & Chadwick, 1968; Spalding et al., 1969a). The microstructure of a eutectic Al–$CuAl_2$ aged for several months is shown in the electron micrograph of Fig. 63. By carefully tilting the specimen it is possible to orientate one of the lamellar interfaces parallel to the direction of incidence of the electron beam. This procedure is necessary to avoid the possible superposition of energy losses from both phases (Cundy, Metherell, & Whelan, 1968). Measurements of the energy losses of electrons contributing to the image at points lying in a line, selected by the analyzer entrance slit, perpendicular to the interface boundary are given in Fig. 64. In region I, which lies in the Al-rich phase and extends over almost all of this phase, the energy loss is constant and has a value of 15.25 ± 0.03 eV, the error quoted being the standard deviation of a number of measurements of the mean plasmon loss. In region II, which extends along the phase boundary and has a width ~1000 Å, the loss varies from 15.25 ± 0.03 eV down to 14.97 ± 0.03 eV. Region III, which represents the interface of the two phases, is 100 Å wide. This width allows for a possible misorientation of $\pm 3°$ in aligning the phase boundary parallel to the incident direction of the electron beam, and in this region the loss is 15.58 ± 0.09 eV. The loss in region IV, which is the θ phase region of the alloy, is constant at 16.34 ± 0.09 eV. The smaller error associated with the loss measurements in the Al-rich phase is due to the fact that the energy loss is more sharply defined (Fig. 65) in this region than in the θ phase region.

The solute concentration profile in the Al-rich phase, obtained from the loss measurements and the calibration curve of Figs. 62 and 64 respectively, is given in Fig. 66. At distances greater than ~100 Å from the

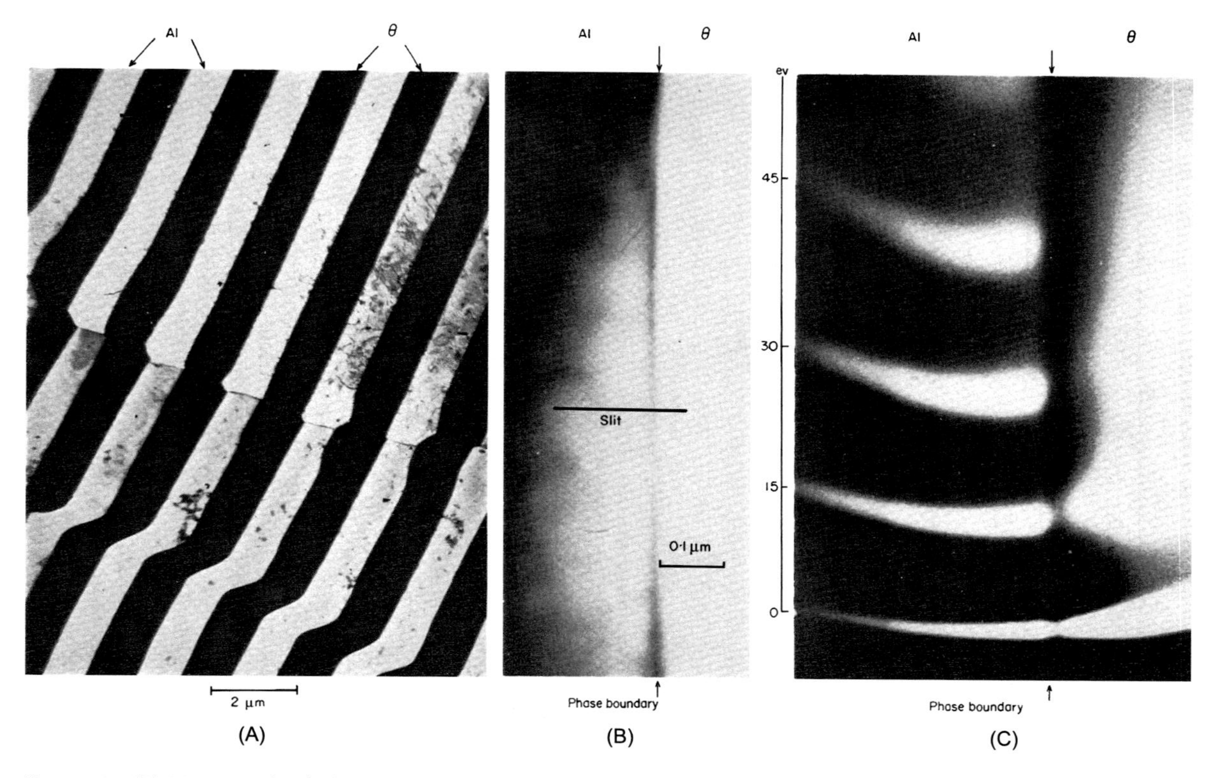

Figure 63 (A) Micrograph of Al–$CuAl_2$ eutectic alloy showing the lamellar substructure of the specimen. The light regions are the Al-rich lamellae (courtesy of F.D. Lemkey and W. Tice). (B) Region of an Al–$CuAl_2$ specimen near the phase boundary (after Spalding, Villagrana, & Chadwick, 1969b). (C) Loss spectrum taken from the line indicated in (B) (after Spalding, 1970).

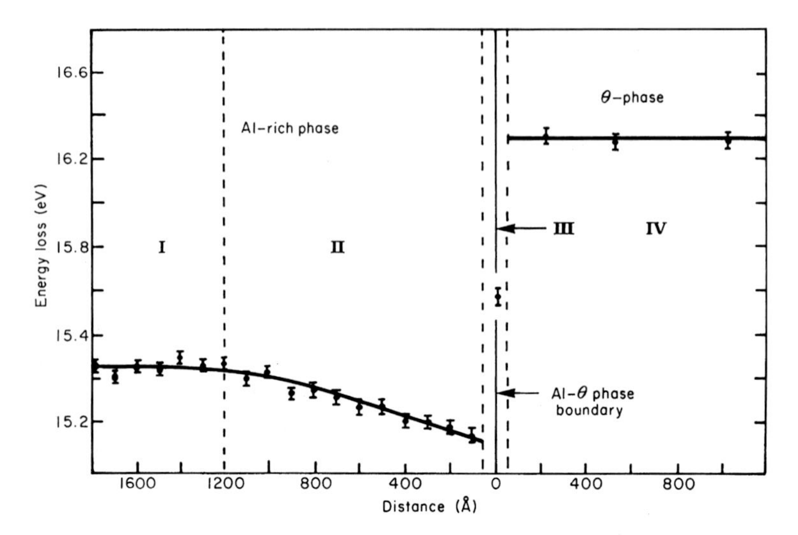

Figure 64 Variation of the mean loss (*P*, Fig. 60) as a function of distance from the phase boundary in an aged specimen of Al–$CuAl_2$ eutectic alloy (after Spalding et al., 1969b, courtesy of *The Philosophical Magazine*).

phase boundary, the solute concentration (~1.5%) is much lower than that expected from eutectic growth theory (~6%, see for example Chadwick, 1963). Similarly the increasing concentration in the vicinity of the interphase boundary is also contrary to that predicted by theory. Suggestions for these discrepancies are given in the paper by Spalding et al. (1969b) to which the interested reader is referred. At the phase boundary the incident electrons are tangential to the interface and a surface plasmon loss (see Raether, 1965) of theoretical value 15.68 eV is expected. This agrees reasonably well with the experimental value of 15.58 eV. The composition range of the $CuAl_2$ phase region predicted by eutectic growth theory is very small (~1%) and this is confirmed by the constancy of the energy losses observed in this region.

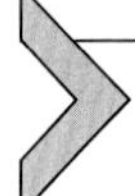

4. ENERGY SELECTING ELECTRON MICROSCOPES

4.1 Instrumentation

The first energy selecting electron microscope to be described here utilizes a cylindrical electrostatic analyzer as the dispersive unit (Watanabe, 1964; Watanabe & Uyeda, 1962a, 1962b). An outline of this instrument, which operates at a beam potential of 100 kV, is given in Fig. 67. The entrance

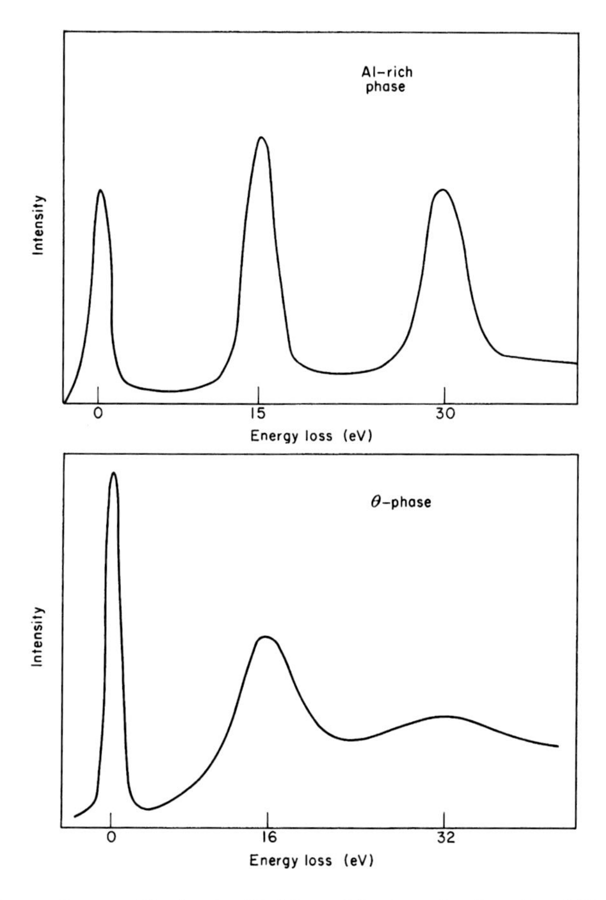

Figure 65 Energy loss profiles in the Al-rich and θ phase (regions I and IV of Fig. 64) of the Al–$CuAl_2$ eutectic alloy (after Spalding et al., 1969b, courtesy of *The Philosophical Magazine*).

aperture of a Möllenstedt energy analyzer is placed in the image plane of the intermediate lens of the microscope. The image formed in this plane is swept backwards and forwards in a direction perpendicular to the long dimension of the analyzer entrance slit by a scan coil placed between the objective and intermediate lenses. A second slit selects one of the loss components of the beam passing through the dispersive system, and this component is scanned by a second coil worked in synchronism with the sweep of the image formed by the intermediate lens. After magnification by the projector lens an energy selected image is therefore swept out on the final image plane of the microscope.

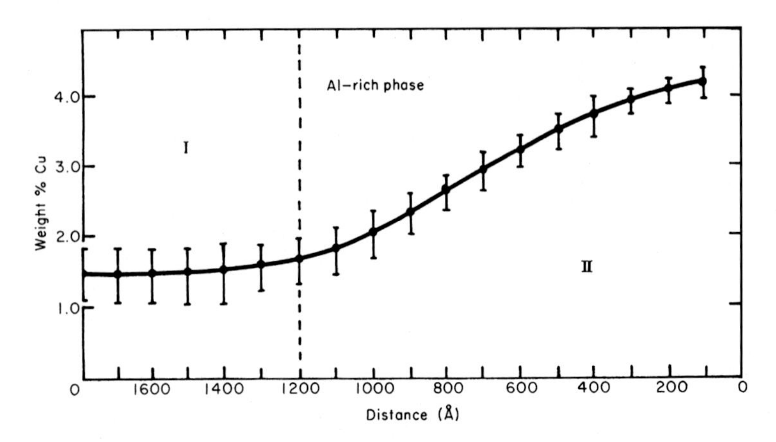

Figure 66 Concentration profile in the Al-rich phase of the Al–$CuAl_2$ eutectic alloy (after Spalding et al., 1969b, courtesy of *The Philosophical Magazine*).

A cylindrical electrostatic lens is placed between the analyzer and the projector lens to eliminate the curvature of the loss line selected to form the filtered image. The reason for this curvature is discussed in Section 3.2, and although there is no necessity for correcting this curvature in the energy analyzing electron microscope, it will, if left uncorrected in the energy selecting microscope, produce a first order distortion in the filtered image. Any curvature of the selected loss line can be eliminated by either translating the cylindrical lens into a suitable off-axis position, or by rotating the lens about an axis parallel to the axes of the cylindrical electrodes.

The energy bandwidth of the filtered image is limited primarily by the energy spread ΔE of the beam incident on the specimen. This spread is usually of the order of 1 or 2 eV at 100 kV and is much larger than the ultimate resolution that can be obtained with the electrostatic analyzer (Section 2.1). It is obviously desirable to obtain a bandwidth which is equal to the energy spread of the incident beam and this is achieved if the width of the energy selecting slit is made equal to the product of the spread ΔE and the dispersion of the analyzer.

An energy selecting electron microscope operating at beam potentials in the range 50 kV to 100 kV and which does not involve the use of scanning techniques has been described in the literature by Castaing and Henry (1962, 1963, 1964) and Henry (1964). This instrument, which is shown in outline in Fig. 68, utilizes the magnetic prism and electrostatic mirror analyzer described in Section 2.3. The cross-over and image plane of a first intermediate lens are made coincident with the real stigmatic point

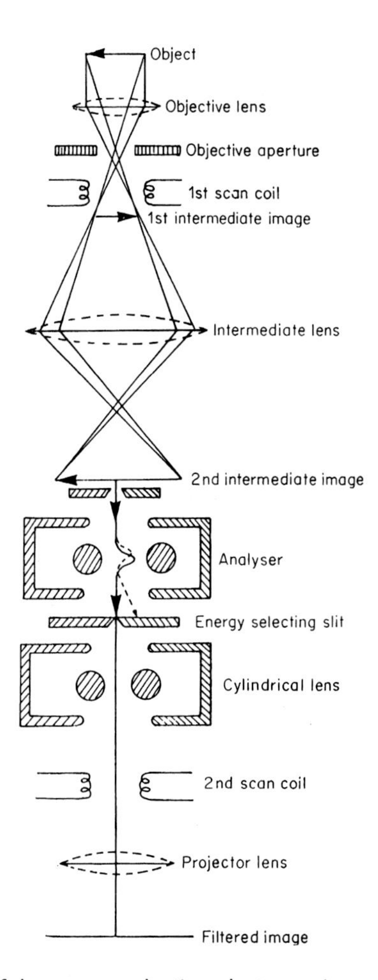

Figure 67 An outline of the energy selecting electron microscope which employs the cylindrical electrostatic analyzer (Section 2.1) as the dispersive unit.

R_1 (Section 2.3) of the magnetic prism and the achromatic plane containing the virtual stigmatic point V_1 respectively. An achromatic image is produced by the dispersive system in the plane containing the virtual stigmatic point V_3 and electrons of different energies are brought to different cross-over points in the plane containing the real stigmatic point R_3. If an aperture is used to select one of the cross-over points produced by a given loss component of the beam, a filtered image can be produced on the final screen by focusing the second intermediate and projector lenses on the achromatic plane containing the point V_3. Alternatively, by removing the

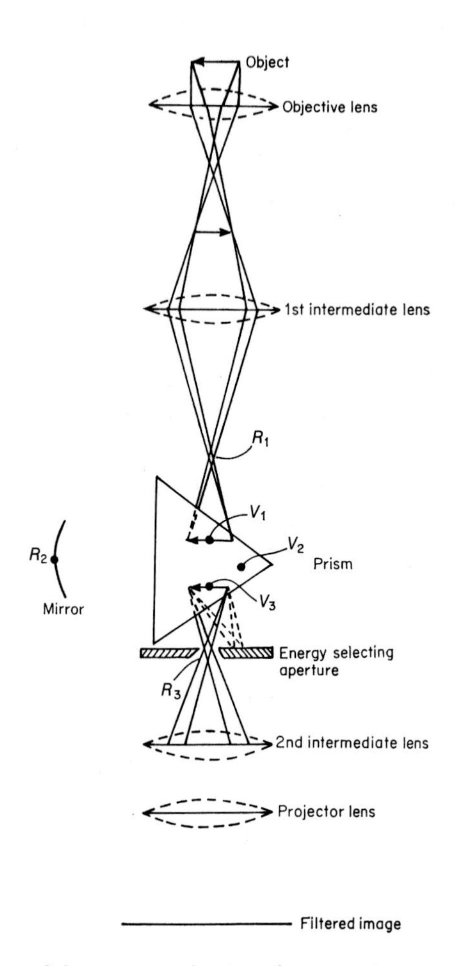

Figure 68 An outline of the energy selecting electron microscope which utilizes the mirror-prism device (Section 2.3).

energy selecting aperture, a conventional unfiltered image of the specimen can be produced on the final image screen of the microscope.

The energy resolution of the device is determined primarily by the size of the cross-over produced at the real stigmatic point R_1. If the diameter of the cross-over at R_1 is d, then since the magnification of the dispersive system is unity, the energy resolution is given by $\Delta E_r = d/D$ where D is the dispersion of the analyzer. The design dimensions of the prism used by Castaing and Henry are (Section 2.3) $\epsilon = 45°$, $a = 4$ cm, $\zeta_1 = -8$ cm, $\zeta_1' = 5$ cm, and for this geometry the dispersion at the plane containing the point R_3 is 2 μm/V for a beam potential of 100 kV. The focal length of

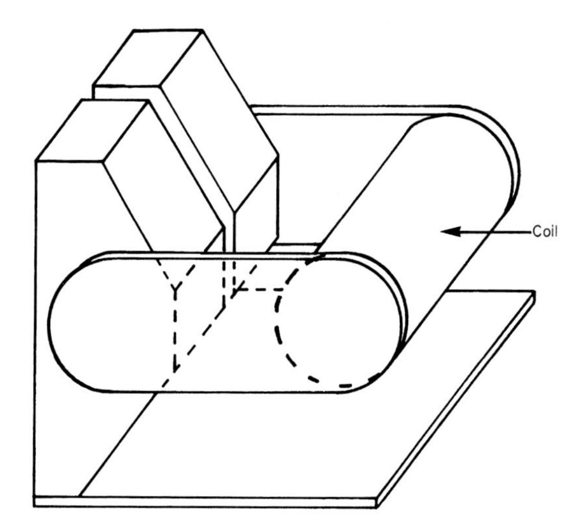

Figure 69 The magnetic prism (after Henry, 1964).

the microscope objective lens is $f_1 = 3.2$ mm and for an objective aperture semi-angle $\theta = 10^{-3}$ rad the cross-over diameter produced by the objective is $d_1 = 2f_1\theta = 6.4$ μm. If the analyzer is placed immediately after the objective lens the resolution obtained is ~3 eV at 100 kV. To obtain a smaller resolution requires a reduction in the cross-over diameter and this can be achieved by placing the dispersive system after the first intermediate lens. In the instrument described by Castaing and Henry this lens is situated at a distance $l = 160$ mm from the objective and has a focal length f_2 which is variable between 15 and 25 mm. The diameter of the cross-over produced by the first intermediate lens is $d_2 = 2f_1\theta f_2/l$, and for the values of f_2 given above, this gives 0.6 μm $< d_2 <$ 1 μm. The resolution limited by the cross-over size is therefore ~0.5 eV and in practice this means that the resolution is limited mainly by the spread in energy of the incident beam which is typically of the order of 1 or 2 eV at 100 kV.

The magnetic prism, a sketch of which is given in Fig. 69, has a pole-piece gap of 2 mm and is excited by a coil consisting of 5000 turns of copper wire of diameter 0.13 mm wound on an Araldite core. At a beam potential of 100 kV an excitation of about 43 amp. turns is required to operate the prism and the resulting heat dissipation (~0.5 W) is sufficiently small to allow the coil to be run in vacuum. Mechanical translation in directions perpendicular and parallel to the axis of the coil allows alignment of the prism with respect to the axis of the electrostatic mirror.

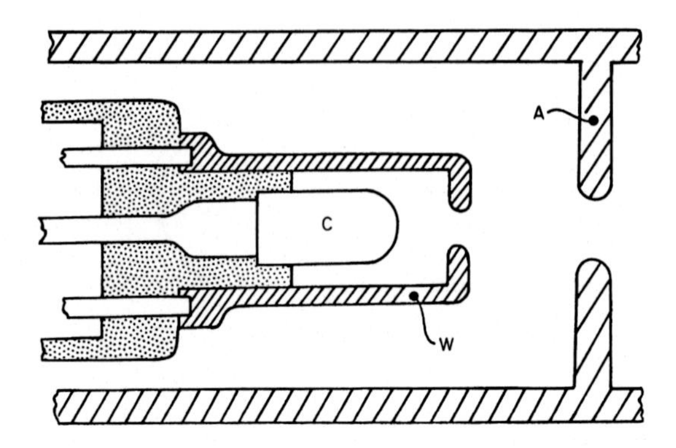

Figure 70 The electrostatic mirror. A: anode, C: cathode, W: Wehnelt cylinder (after Henry, 1964).

The mirror consists of a modified electron gun, in which the filament has been replaced by a cylindrical cathode *C* (Fig. 70). The cathode is biased slightly negative with respect to the beam potential and the Wehnelt cylinder is biased negative with respect to the cathode. The electrode system therefore has a field distribution similar to that of an immersion lens. Adjustment of the position of the pole of the mirror is effected by mechanical movement of the gun and the focal length is adjusted by altering the bias of the cathode and Wehnelt cylinders.

4.2 Applications

Most of the experimental work involving the use of energy selecting electron microscopes has been directed at investigating the preservation of image contrast by inelastically scattered electrons. By displacing the objective aperture of a conventional electron microscope so that no Bragg scattered electrons contribute to the image, Kamiya and Uyeda (1961) demonstrated (Fig. 71) that inelastic scattering does not destroy image contrast. In this simple experiment the energy losses of the electrons are unknown and it is therefore impossible to discriminate between different types of inelastic scattering process. Using the energy selecting electron microscope described in Section 4.1, Watanabe and Uyeda (1962a, 1962b) were able to show that electrons which lose energy by plasmon excitation can be used to form filtered images which exhibit contrast effects similar to those observed in the ordinary image (Fig. 72). This result was given theoretical justification by Fujimoto and Kainuma (1961, 1963), Howie (1962, 1963), and Fukuhara (1963). A further series of experiments (Castaing, Hili, &

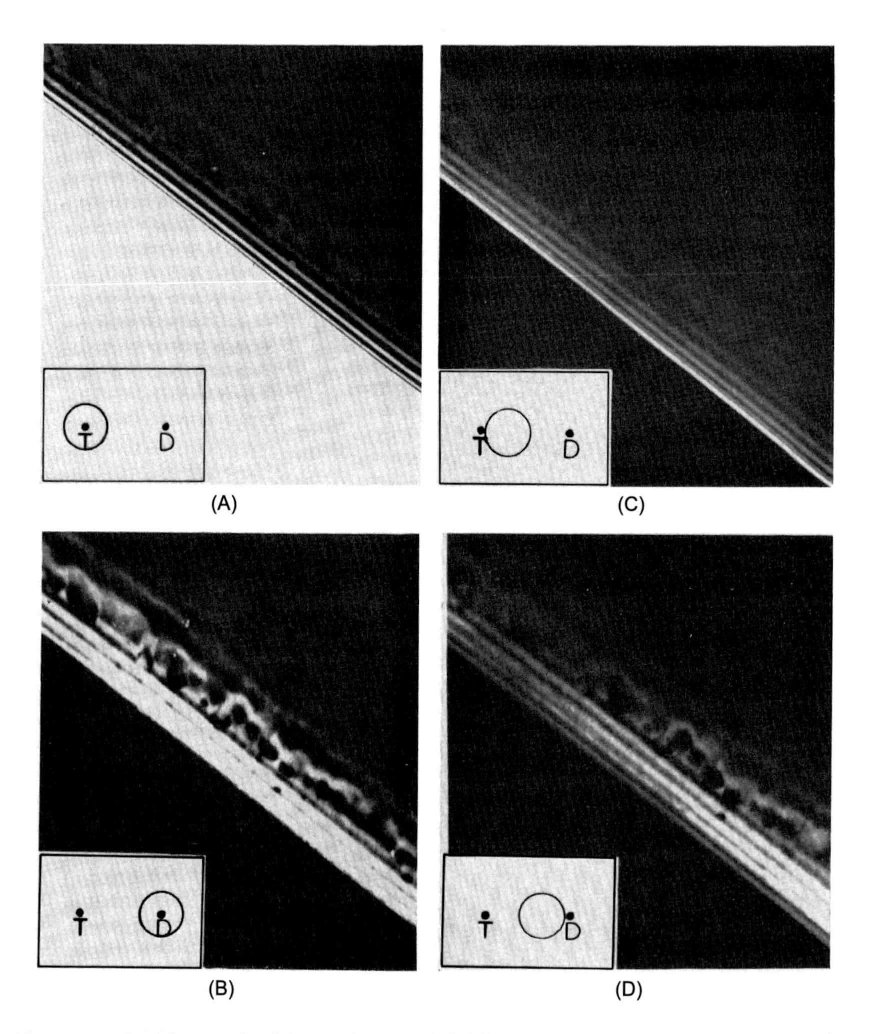

Figure 71 (A) The bright field and (B) dark field images of a specimen of aluminum. (C) The image obtained by displacing the objective aperture so as to exclude the Bragg scattered electrons. The position of the aperture is indicated in the inset where T represents the central spot and D the diffracted spot of the diffraction pattern. (D) The image obtained when the objective aperture is displaced in the vicinity of the diffracted spot D. In (A) and (B) both elastically and inelastically scattered electrons contribute to the contrast observed, whereas in (C) and (D) only the inelastically scattered electrons contribute to contrast.

Henry, 1966a, 1966b; Cundy et al., 1969; Cundy et al., 1966) have shown that all electrons with losses lying between about 1 eV and 100 eV also appear to preserve image contrast.

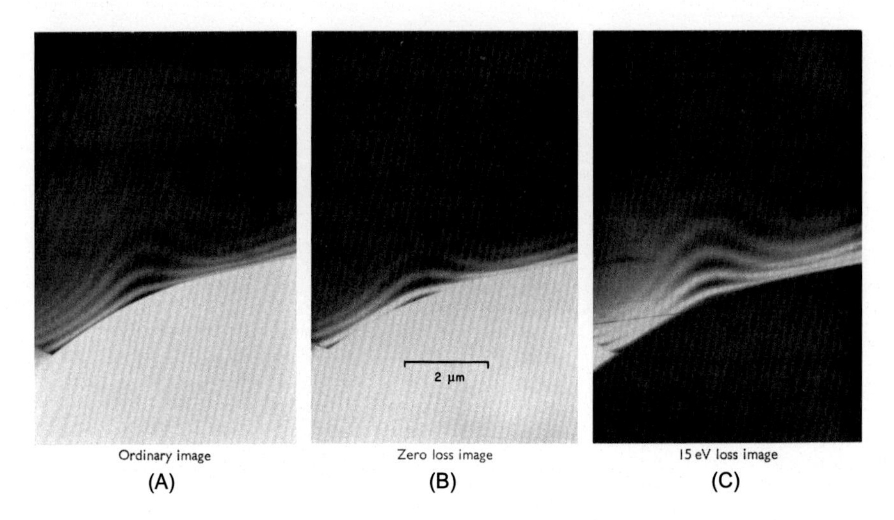

Figure 72 Examples of energy selected images of Al at 100 kV. (A) The ordinary image. (B) The zero-loss image and (C) the 15 eV loss image showing the preservation of contrast by plasmon scattered electrons (courtesy of Dr. H. Watanabe).

The question remains as to whether or not thermal diffuse scattering also preserves contrast. It is not possible with present instrumentation to resolve the energy losses of electrons which have excited phonons, since these lie within the energy spread of the incident beam. An indirect experiment, in which the objective aperture is displaced so that no Bragg beams contribute to the image, can however determine whether or not contrast is preserved by inelastic scattering processes producing energy losses in the range 0–1 eV. Most of the electrons contributing to this loss range can be expected to have been inelastically scattered by the excitation of phonons, since the displaced aperture selects only those electrons in the diffuse background surrounding the Bragg spots of the diffraction pattern. Filtered images of aluminum specimens obtained with a displaced objective aperture have been published by Castaing et al. (1966a, 1966b), and these show weak contrast effects in the vicinity of thickness fringes and bend contours. Reduced contrast was also observed by Cundy et al. (1967a), at thickness fringes in aluminum, but these authors pointed out that since the contrast was so weak there was some uncertainty as to whether it was due to phonon scattering or due to electrons entering the objective aperture after elastic scattering in oxide or contamination layers present on the surface of the aluminum specimens. In a further series of experiments (Castaing, Henoc, & Henry, 1968; Castaing, Henoc, Henry, & Natta, 1967) great care was taken to en-

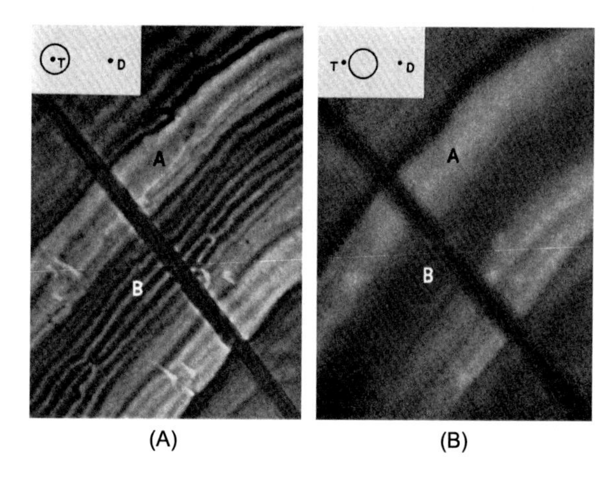

Figure 73 Filtered images of a specimen of Au obtained by selecting electrons with energy losses less than 1 eV. The positions of the objective aperture used to form these images are shown in the insets. In (A) the strong fringe contrast at A and B is due almost entirely to Bragg scattered electrons. In (B) which is the image formed by the inelastic component with losses less than 1 eV, the fringe contrast at A and B has almost entirely disappeared (after Hili, 1967, courtesy of Professor R. Castaing, Dr. L. Henry and *Journal de Microscopie*).

sure that the specimens examined were free of surface films. This was achieved by using specimens of gold, to eliminate oxide films, mounted in an anti-contamination stage, to reduce the deposition rate of carbon on the specimen surfaces to a negligible level. Even under these stringent conditions weak contrast effects (Fig. 73) are still observed, provided that sufficiently small objective apertures are used, indicating that a proportion of the phonon scattered electrons preserve image contrast. In an experiment reported by Cundy et al. (1969), however, no discernible contrast at stacking fault images could be observed in the inelastic component of the 0–1 eV loss. These authors used an objective aperture somewhat larger than those employed by Castaing and his colleagues and it is possible that the discrepancy between the two sets of results lies in a destruction of contrast by diffuse scattering, as discussed by Metherell (1967b) for the case of plasmon scattering.

A study of the preservation of contrast by plasmon scattered electrons in out of focus images has been made by Spalding (1970) using an energy analyzing electron microscope. A through focal series of electron micrographs of a wedge shaped specimen of aluminum is given in Fig. 74A, B, C, and D. The loss spectra of electrons arriving at the lines indicated on each image,

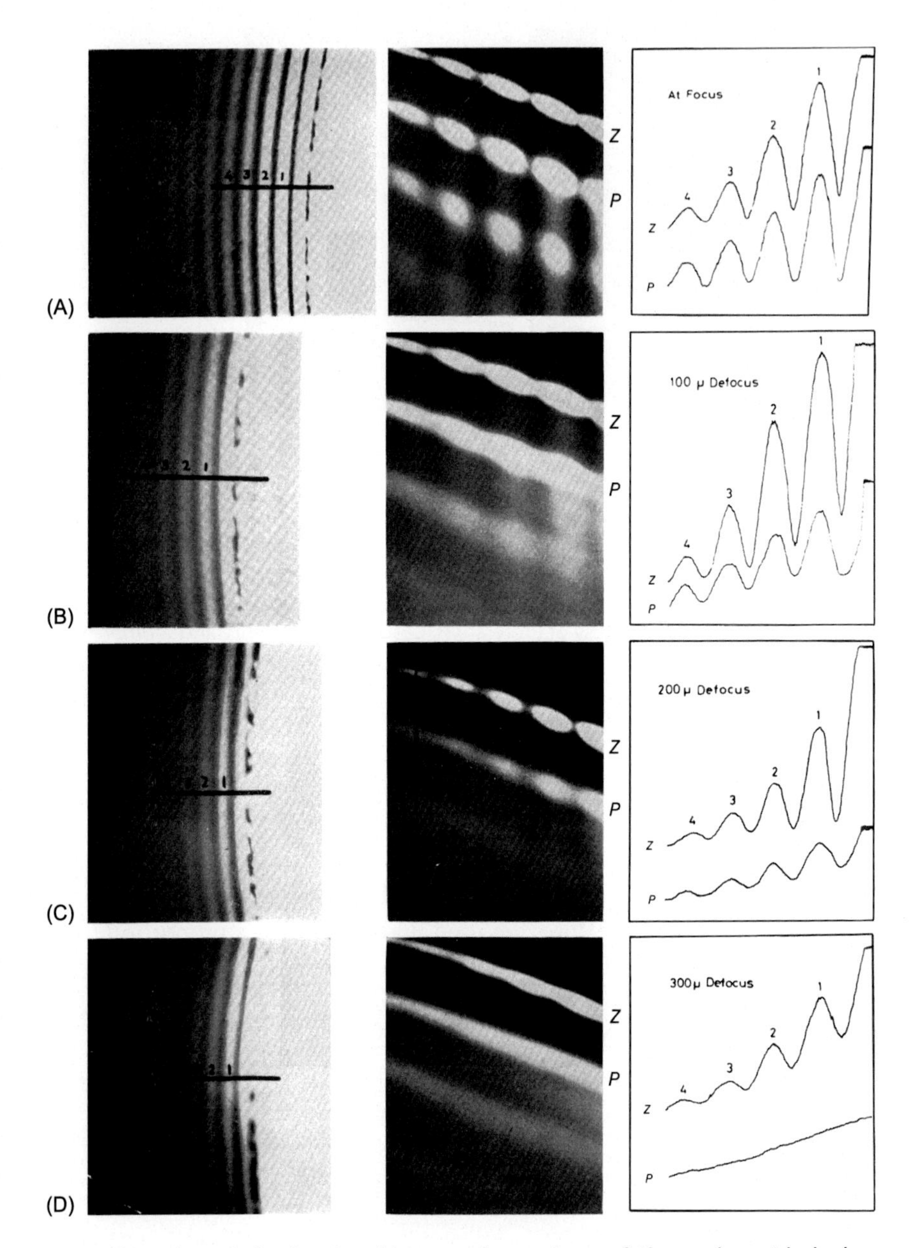

Figure 74 A through focal series of images of a specimen of Al, together with the loss spectra from the lines indicated in the micrographs. Microdensitometer traces of the zero loss (*Z*) and plasmon loss (*P*) lines are also given, (A) at focus, (B) 100 μm defocus, (C) 200 μm defocus, (d) 300 μm defocus. The contrast in the plasmon loss line disappears at a defocus of 300 μm (after Spalding, 1970).

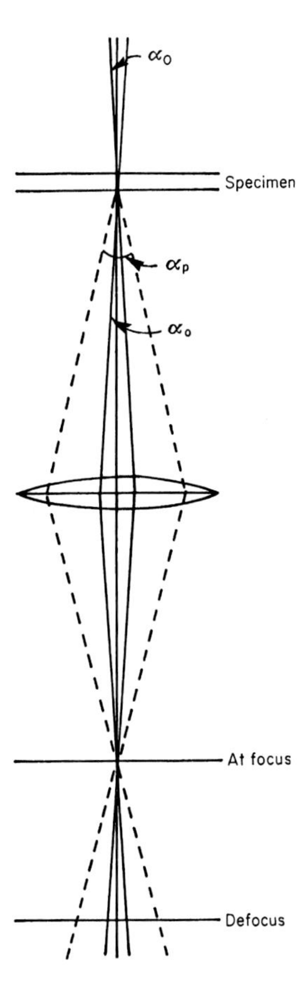

Figure 75 Diagram illustrating the mechanism responsible for the loss of contrast observed in the through focal series of micrographs of Fig. 74.

together with microdensitometer traces which show the thickness fringe intensity variations along the zero loss (Z) and plasmon loss lines (P), are also given in this figure. At a defocus of 300 μm (Fig. 74D) the contrast due to the plasmon scattered electrons has disappeared completely, whereas some contrast is still observable in the zero loss electrons. The reason for this difference in behavior of the elastically and inelastically scattered electrons is almost certainly due to the greater divergence angle associated with the latter on emergence from the specimen. The angle of collimation α_0 (Fig. 75) of the incident beam is typically $\sim 10^{-4}$ rad, whereas the mean angle of scatter α_p for 100 kV electrons which have lost energy by plasmon excitation is $\sim 10^{-3}$ rad. The image formed by the plasmon scattered elec-

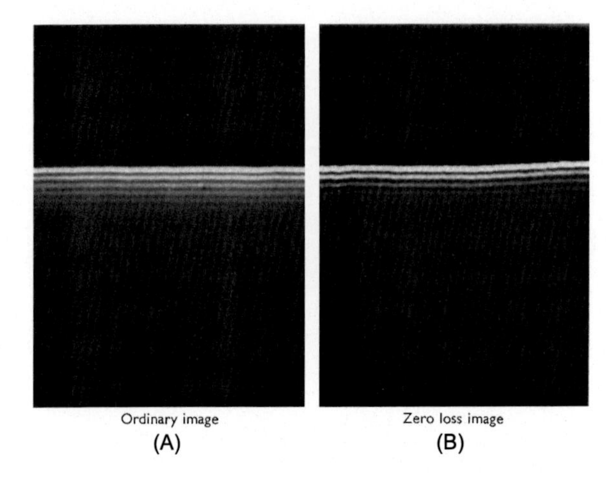

Figure 76 (A) The ordinary image and (B) the zero loss image of an MgO specimen at 75 kV. Exclusion of the inelastically scattered component of the electron beam changes the mean absorption coefficient for the ordinary image from μ_0 to $\mu_0 + 1/\lambda$ for the zero loss image, where λ is the mean free path for inelastic scattering (courtesy of Dr. H. Watanabe).

trons therefore goes out of focus more rapidly than the zero loss electrons as defocus proceeds.

The effect of inelastic scattering on absorption contrast has been studied by Watanabe (1966). In his experiments the mean absorption coefficients of the ordinary image (Fig. 76A) and the zero-loss image (Fig. 76B) of an MgO specimen were obtained from measurements of the intensity profiles of thickness fringes. A discussion of his results can be found in papers by Cundy, Metherell, and Whelan (1967b) and Metherell (1967c) to which the interested reader is referred.

As a microanalyzer the energy selecting electron microscope appears at present to be of somewhat more restricted application than the energy analyzing electron microscope; the reason being that the bandwidth of filtered images is limited by the energy spread of the incident beam ($\sim$1 eV) whereas the microanalysis experiments, described in Section 3.2, involving the use of an energy analyzing electron microscope, rely on the measurement of the mean plasmon energy and with care random errors associated with these measurements can be reduced $\pm$0.03 eV (see for example Spalding et al. (1969a) for a discussion of errors). It is therefore possible, for example, to detect differences in the solute concentration in the α phase of the Al–Mg alloy amounting to about 1%. Using filtered images with a bandwidth of 1 eV, it would be impossible to detect solute

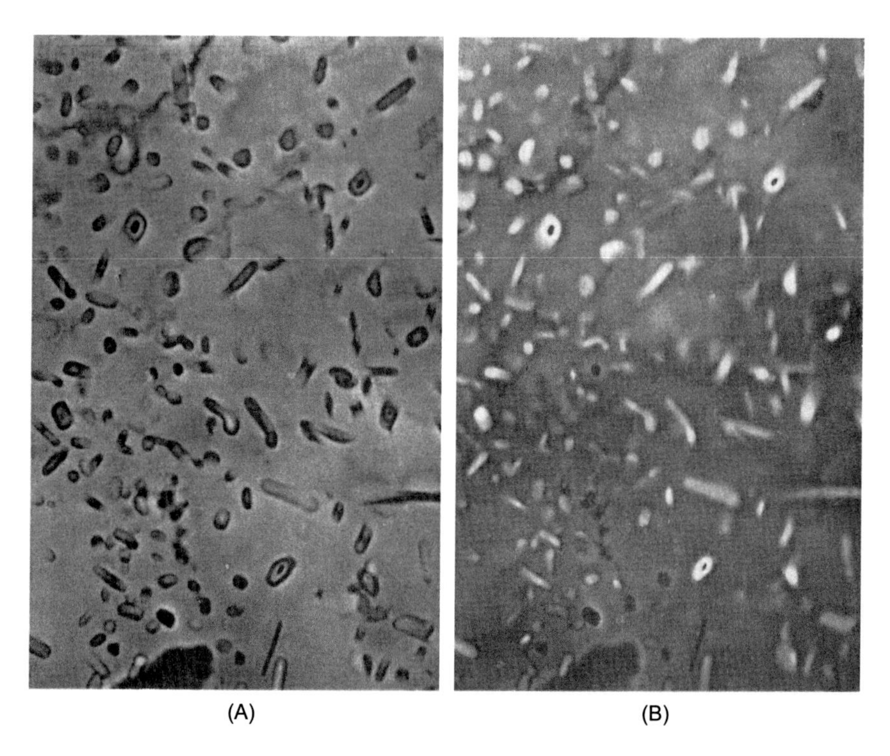

Figure 77 Filtered images obtained with a specimen of Al-7.6% Zn-2.6% Mg alloy. (A) The 4.6 ± 1 eV image showing the η phase precipitates, which appear dark against a light background. (B) The 22.5 ± 2.5 eV image, in which the η phase precipitates appear light against the darker background (after Hili, 1966, courtesy of Professor R. Castaing, Dr. L. Henry and *Journal de Microscopie*).

concentration differences in the same alloy of much less than about 10%. The energy selecting microscope can, however, be used to identify different phases of a precipitated alloy system, provided that the energy losses associated with the different phases are sharply defined and separated by more than about 1 eV. In this situation it is possible to form a filtered image using the energy loss characteristic of one of the phases. An example of an alloy which meets these requirements is Al-7.6% Zn-2.6% Mg which, when suitably heat-treated, forms η phase precipitates (composition $MgZn_2$) imbedded in the solid solution phase of the system. The loss peak of the η phase occurs at ~22 eV whereas that of the matrix material occurs at ~14.5 eV, and the filtered images obtained by selecting the 14.5 ± 1 eV and 22 ± 2 eV losses are shown in Figs. 77A and B respectively. In Fig. 77A the η phase precipitates appear dark against matrix material and in Fig. 77B the reverse occurs. Using this technique Castaing and his colleagues found that it was possible to identify η phase precipitates as small as 100 Å across.

ACKNOWLEDGMENTS

The author is indebted to Drs. K.T. Considine, P.W. Hawkes, and K.C.A. Smith for useful discussions and to Professor R. Castaing, Dr. L. Henry, and Dr. H. Watanabe for allowing him to reproduce examples of filtered images obtained with their energy selecting electron microscopes.

REFERENCES

Archard, G. D. (1954). *British Journal of Applied Physics*, *5*, 179, 395.

Castaing, R., & Henry, L. (1962). *Comptes rendus hebdomadaires des séances de l'Académie des Sciences (Paris)*, *255*, 76.

Castaing, R., & Henry, L. (1963). *Journal de Microscopie*, *2*, 5.

Castaing, R., & Henry, L. (1964). *Journal de Microscopie*, *3*, 133.

Castaing, R., Henoc, P., & Henry, L. (1968). In *Proc. 4th European reg. conf. elect. microsc., Rome I*, 285.

Castaing, R., Henoc, P., Henry, L., & Natta, M. (1967). *Comptes rendus hebdomadaires des séances de l'Académie des Sciences (Paris)*, *265*, 1293.

Castaing, R., Hili, A. El, & Henry, L. (1966a). *Comptes rendus hebdomadaires des séances de l'Académie des Sciences (Paris)*, *262*, 169.

Castaing, R., Hili, A. El, & Henry, L. (1966b). *Comptes rendus hebdomadaires des séances de l'Académie des Sciences (Paris)*, *262*, 1051.

Chadwick, G. A. (1963). *Progress in Materials Science*, *12*(2).

Considine, K. T. (1970). Ph.D. Thesis, Univ. of Cambridge.

Considine, K. T., & Smith, K. C. A. (1968). In *Proc. 4th European reg. conf. elect. microsc., Rome I*, 329.

Cook, R. F., & Howie, A. (1969). *Philosophical Magazine*, *20*, 641.

Cotte, M. (1938). *Annales de Physique (Paris)*, *10*, 333.

Cundy, S. L. (1968). Ph.D. Thesis, Univ. of Cambridge.

Cundy, S. L., Howie, A., & Valdre, U. (1969). *Philosophical Magazine*, *20*, 147.

Cundy, S. L., Metherell, A. J. F., & Whelan, M. J. (1966). *Journal of Scientific Instruments*, *43*, 712.

Cundy, S. L., Metherell, A. J. F., & Whelan, M. J. (1967a). *Philosophical Magazine*, *15*, 623.

Cundy, S. L., Metherell, A. J. F., & Whelan, M. J. (1967b). *Physics Letters A*, *24*, 120.

Cundy, S. L., Metherell, A. J. F., & Whelan, M. J. (1968). *Philosophical Magazine*, *17*, 141.

Cundy, S. L., Metherell, A. J. F., Whelan, M. J., Unwin, P. N. T., & Nicholson, R. B. (1968). *Proceedings of the Royal Society. Series A*, *307*, 267.

Dietrich, W. (1958). *Zeitschrift für Physik*, *151*, 519.

Ferrell, R. A. (1956). *Physical Review*, *101*, 554.

Fujimoto, F., & Kainuma, Y. (1961). *Journal of the Physical Society of Japan, Suppl. BII*, *17*, 140.

Fujimoto, F., & Kainuma, Y. (1963). *Journal of the Physical Society of Japan*, *18*, 1792.

Fukuhara, A. (1963). *Journal of the Physical Society of Japan*, *18*, 496.

Gauthe, B. (1954). *Comptes rendus hebdomadaires des séances de l'Académie des Sciences*, *239*, 399.

Hansen, M. H. (1958). *Constitution of binary alloys* (2nd ed.). New York: McGraw-Hill.
Hennequin, J. F. (1960). Diplôme d'etudes supérieures, Univ. of Paris.
Henry, L. (1964). Doctorate Thesis, Univ. of Paris.
Hili, A. El (1966). *Journal de Microscopie, 5*, 669.
Hili, A. El (1967). *Journal de Microscopie, 6*, 725.
Hillier, J. (1943). *Physical Review, 64*, 318.
Howie, A. (1962). In *Proc. 5th int. conf. electron microsc. (Philadelphia), vol. 1*, AA-10.
Howie, A. (1963). *Proceedings of the Royal Society. Series A, 271*, 268.
Ichinokawa, T. (1965). In *Proc. int. conf. electron diff. and crystal defects (Melbourne)*, IN-4.
Ichinokawa, T. (1968). *Japanese Journal of Applied Physics*, 7, 799.
Ichinokawa, T., & Kamiya, Y. (1966). In *Proc. 6th int. conf. electron microsc. (Kyoto), 1*, 89.
Kamiya, Y., & Uyeda, R. (1961). *Journal of the Physical Society of Japan, 16*, 1361.
Kleinn, W. (1954). *Optik, 11*, 226.
Klemperer, O. (1965). *Reports on Progress in Physics, 28*, 77.
Laudet, M. (1953). *Cahiers de Physique, 41*, 72.
Leithäuser, E. (1904). *Annalen der Physik (Leipzig), 15*, 283.
Lenz, F. (1953). *Optik, 10*, 439.
Leonhard, F. (1954). *Zeitschrift für Naturforschung, 9a*, 727.
Lippert, W. (1955). *Optik, 12*, 467.
Marton, L. (1944). *Physical Review, 66*, 159.
Marton, L., & Leder, L. B. (1954). *Physical Review, 94*, 203.
Metherell, A. J. F. (1965). Ph.D. Thesis, Univ. of Cambridge.
Metherell, A. J. F. (1967a). *Optik, 25*, 250.
Metherell, A. J. F. (1967b). *Philosophical Magazine, 15*, 763.
Metherell, A. J. F. (1967c). *Philosophical Magazine, 16*, 1103.
Metherell, A. J. F., & Cook, R. F. (1972). *Optik, 34*, 535–552.
Metherell, A. J. F., Cundy, S. L., & Whelan, M. J. (1965). In *Proc. int. conf. electron diff. and crystal defects (Melbourne)*, IN-3.
Metherell, A. J. F., & Whelan, M. J. (1965). *British Journal of Applied Physics, 16*, 1038.
Metherell, A. J. F., & Whelan, M. J. (1966). *Journal of Applied Physics, 37*, 1737.
Möllenstedt, G. (1949). *Optik, 5*, 499.
Möllenstedt, G. (1952). *Optik, 9*, 473.
Paras, N. (1961). Diplôme d'etudes supérieures, Univ. of Paris.
Raether, H. (1965). *Springer Tracts in Modern Physics, 38*, 84.
Raimes, S. (1967). *The wave mechanics of electrons in metals*. North-Holland Publishing Co.
Septier, A. (1954). *Comptes rendus hebdomadaires des séances de l'Académie des Sciences (Paris), 239*, 402.
Smith, K. C. A., Considine, K. T., & Cosslett, V. E. (1966). In *6th int. conf. elect. microsc. (Kyoto) I*, 99.
Spalding, D. R. (1970). Ph.D. Thesis, Univ. of Cambridge.
Spalding, D. R., Edington, J. W., & Villagrana, R. E. (1969a). *Philosophical Magazine, 20*, 1203.
Spalding, D. R., & Metherell, A. J. F. (1968). *Philosophical Magazine, 18*, 41.
Spalding, D. R., Villagrana, R. E., & Chadwick, G. A. (1968). In *Proc. 4th Europ. reg. conf. elect. microsc. (Rome) I*, 347.
Spalding, D. R., Villagrana, R. E., & Chadwick, G. A. (1969b). *Philosophical Magazine, 20*, 471.
Watanabe, H. (1954). *Journal of the Physical Society of Japan, 9*, 920.

Watanabe, H. (1955). *Journal of the Physical Society of Japan, 10*, 321.
Watanabe, H. (1956). *Journal of the Physical Society of Japan, 11*, 112.
Watanabe, H. (1964). *Japanese Journal of Applied Physics, 3*, 480.
Watanabe, H. (1966). In *Proc. 6th int. conf. elect. microsc. (Kyoto) I*, 63.
Watanabe, H., & Uyeda, R. (1962a). In *Proc. 5th int. cong. elect. microsc. (Philadelphia) I*, A-5.
Watanabe, H., & Uyeda, R. (1962b). *Journal of the Physical Society of Japan, 17*, 569.
Waters, W. E. (1956). Ph.D. Thesis, University of Maryland.
Wien, W. (1897). *Verhandlungen der Deutschen Physikalischen Gesellschaft, 16*, 165.

INDEX

F

G

H

I

L

Printed in the United States
By Bookmasters